Interior Point Approach to Linear, Quadratic and Convex Programming

Mathematics and Its Applications

Managing Editor:

M. HAZEWINKEL
Centre for Mathematics and Computer Science, Amsterdam, The Netherlands

Volume 277

Interior Point Approach to Linear, Quadratic and Convex Programming
Algorithms and Complexity

by

D. den Hertog
Centre for Quantative Methods,
Eindhoven, The Netherlands

SPRINGER SCIENCE+BUSINESS MEDIA, B.V.

A C.I.P. Catalogue record for this book is available from the Library of Congress.

ISBN 978-94-010-4496-7   ISBN 978-94-011-1134-8 (eBook)
DOI 10.1007/978-94-011-1134-8

02-0197-100 ts
Reprinted 1997.

Printed on acid-free paper

Aan mijn ouders

Voor Annemieke en
Geert-Jan

Contents

Acknowledgements

This book was written when the author was working at the Delft University of Technology. During this period I have had the benefit of advice and help from many people.

First I want to thank Dr. ir. C. Roos for his stimulating and enthusiastic collaboration. During the past several years, I had the good fortune to work in his research group dedicated to the study of interior point methods. He always gave freely of his time, whether at his office during our daily meetings or occasionally by telephone late into the evening.

I also wish to acknowledge Dr. T. Terlaky, a veteran of the interior point group, for our abundant collaboration which I always found to be helpful and fruitful. His broad view of the mathematical programming area was of great value for me.

Thanks also go to Prof. dr. F.A. Lootsma who urged me to rearrange my work into a book.

Also the contact with the KSLA (Koninklijke/Shell–Laboratorium, Amsterdam) group in Amsterdam, which was formalized in an official research agreement, was essential. I gratefully acknowledge KSLA for the financial support during the past two years. In particular I want to thank Drs. J.F. Ballintijn for the discussions we had.

I also learned a lot from several guests who visited Delft: Prof. dr. K.M. Anstreicher, Dr. O. Güler, Dr. F. Jarre, Dr. J. Kaliski, and my second promotor Prof. dr. J.–Ph. Vial. Many of these visits resulted in joint publications.

I also have to mention my officemates Jan van Noortwijk and Benjamin Jansen. The discussions with Jan on the role of the Operations Research in practice and the relationship between theory and practice influenced my choice of careers. Benjamin, also a member of the interior point group, read the first version of this book very carefully. The task he had set for himself was to find at least one error on each page. I have to admit that he succeeded. Nevertheless, I take full responsibility for all remaining mistakes in this book.

Furthermore, I thank all other persons in the university who made my stay in Delft pleasant, especially Wim Penninx for his TEX assistance.

I want to thank my parents for all that they have done for me, which is too much to write down. Finally, I sincerely thank my wife Annemieke for all of her love and compassion. She understood the many times I was physically or mentally absent from her. For this, I am in her debt.

Glossary of symbols and notations

x_i	i–th coordinate of the vector x.
x^T	transpose of the vector x.
X	diagonal matrix with the components of x on the diagonal.
e	vector of all 1's of appropriate length, i.e. $e^T = (1, \cdots, 1)^T$.
e_i	i–th unit vector of appropriate length.
I	identity matrix of appropriate dimension.
$\mathbb{R}^n$	n–dimensional Euclidean space.
$\|x\|$	2–norm of the vector x.
$v(n) = O(w(n))$	means that there exists a constant $c > 0$ such that, for large enough n, $v(n) \leq cw(n)$.
$v(n) = \Theta(w(n))$	means that there exist constants $c_1 > 0$ and $c_2 > 0$ such that, for large enough n, $c_1 w(n) \leq v(n) \leq c_2 w(n)$.
$F \preceq G$	means that $G - F$ is positive semi–definite.
L	binary input length of the linear (or quadratic) programming problem.
$(\mathcal{LP})$	primal linear programming problem.
$(\mathcal{LD})$	dual linear programming problem.
$(\mathcal{QP})$	primal quadratic programming problem.
$(\mathcal{QD})$	dual quadratic programming problem.
$(\mathcal{CP})$	primal convex programming problem.
$(\mathcal{CD})$	(Wolfe) dual of the convex programming problem.
z^*	optimal value.
z or z_l	a lower bound for the optimal value.
z_u	an upper bound for the optimal value.
μ	barrier parameter.
μ_0	initial barrier parameter.
z^0	initial lower bound for the optimal value.
ϕ_B	logarithmic barrier function.
ϕ_D	distance function of Huard.
ϕ_K	Karmarkar's potential function.
ϕ_{TY}	primal–dual potential function of Todd and Ye.
ϕ_M	multiplicative potential function.
ϕ_{SM}	symmetric multiplicative potential function.
$x(\mu), y(\mu)$	minimizing point for ϕ_B (primal, dual).
$x(z), y(z)$	minimizing point for ϕ_D (primal, dual).
$\delta(y, \mu)$	distance measure to $y(\mu)$.

$\delta(x,\mu)$	distance measure to $x(\mu)$.
$\sigma(y,z)$	distance measure to $y(z)$.
P_B	orthogonal projection onto the null–space of the matrix B.
f_0	objective function of $(\mathcal{CP})$.
$f_i,\ i=1,\cdots,n$	constraint functions of $(\mathcal{CP})$.
p	search direction.
g	gradient of ϕ_B or ϕ_D.
H_i	Hessian of $f_i(y)$.
H	Hessian of ϕ_B or ϕ_D.
κ	self–concordance constant.
$\mathcal{F}$	feasible region of problem $(\mathcal{CP})$.
$\mathcal{F}^0$	interior of $\mathcal{F}$.
$\mathcal{F}_z$	intersection of $\mathcal{F}$ and the level set $f_0(y) \geq z$.
$\mathcal{F}_z^0$	interior of $\mathcal{F}_z$.
ϵ	desired accuracy for the final solution.
θ	updating factor for μ or z.
α	steplength.

Chapter 1

Introduction of IPMs

In this chapter we describe and motivate the subject of this book. We start by describing the previous history of interior point methods. Then, we recall some complexity issues. Finally, after describing the recent history of interior point methods, we explain the main goal and contribution of this book.

1.1 Prelude

The classical methods for solving the nonlinear constrained optimization problem

$$\max \left\{ f_0(y) : \ y \in \mathbb{R}^m, \ f_i(y) \le 0, \ i = 1, \cdots, n \right\}$$

can roughly be distinguished into two classes: *interior* and *exterior point* methods. The interior point methods operate in the interior of the feasible region, while exterior point methods try to obtain a solution from outside the feasible region. So, *interior point methods* (IPMs) are not new. In fact, already in the 1960s some interior point methods for nonlinear programming, e.g. barrier methods, were proposed and analyzed.

Linear programming (LP) was always considered as distinct from nonlinear programming because of its combinatorial character. Dantzig's simplex method has always been the unbeaten tool for solving the linear programming problem. The simplex method explicitly uses the combinatorial structure: it obtains an optimum by moving along the edges of the feasible region.

In the 1960s and 1970s, due to the fast computer developments, researchers became interested in the *complexity* of methods. A method was called *polynomial* if the number of arithmetic operations required by the method to solve the problem, is bounded from above by a polynomial in the problem size. Such a polynomial method was considered to be an efficient method.

In 1972, Klee and Minty [75] gave a linear programming example for which the simplex method, using a certain pivot rule, needs $2^n - 1$ iterations, whereas the problem has n variables and $2n$ constraints. 'Exponential examples' for other simplex pivot rules were soon constructed by other researchers[1]. The practical performance of the simplex method however is very good; the simplex method was considered to be the champion forever. Nevertheless, the question arose if it is possible to construct a polynomial method for LP.

In 1979, Khachian [73] published the first polynomial algorithm for LP. This ellipsoid method was based on some nonlinear programming techniques developed by Shor [134] and Yudin and Nemirovsky [168]. Contrary to the simplex method, the ellipsoid method does not use the combinatorial structure of LP. Although this first polynomial method has many important theoretical consequences, it soon appeared that in practice this method is hopelessly slow. Consequently, the quest changed to find a polynomial method for LP also being efficient in practice.

Karmarkar [72] answered this question in an epoch–making work in 1984. He proposed a new polynomial method and claimed that this so–called *projective method* could solve large scale linear programming problems as much as 100 times faster than the simplex method. This claim was received with much scepticism. Nowadays, it has become clear that Karmarkar's work has opened a new research field, the field of interior point methods, which has yielded many theoretical and practical jewels.

The last eight years much research has been done in this field, both theoretical and practical. Inspired by Karmarkar's publication many other (polynomial) methods were proposed by researchers, sometimes containing new ideas sometimes containing only slight modifications or embellishments. Through all these developments one could not see the wood for the trees for a long time. Numerical results indeed have shown that some interior point methods can solve large linear programming problems faster than the simplex method.

Another important property of IPMs is that some of these methods can also be applied to nonlinear optimization problems. In fact, some of these methods were developed for these problems in the 1960s already. This leaded to a rebirth of some old methods: the *logarithmic barrier function method* of Frisch [37], [38], and Fiacco and McCormick [35], the *center method* of Huard [60], and the *affine scaling method* of Dikin [31].

In classical barrier methods the objective function and the constraints are combined in a composite function, called the barrier function, such that it contains a singularity at the boundary. Because of this singularity, all iterates will remain strictly feasible while approaching the optimal solution. The logarithmic barrier function was originally introduced by Frisch for linear programming [37] and for convex programming [38]. However, he apparently did not minimize this function sequentially, but only used the gradient of it to steer away from the boundary. Parisot [121] further developed the logarithmic barrier method and devised sequential unconstrained minimization techniques for linear programming.

---

[1]It is still an open question whether there exists a polynomial–time simplex pivot rule for linear programming.

Later on Fiacco and McCormick [35] and Lootsma [86] gave a treatment of the method as a tool for solving nonlinear problems. In this method the logarithmic barrier function is minimized exactly for decreasing values of the barrier parameter. These minima form the so-called central path of the problem, which is a smooth curve in the interior of the feasible region and ends in an optimal solution. Nice proofs of asymptotic convergence have been obtained for this method; see also Fiacco and McCormick [35]. Polak [125] studied more implementable algorithms in which the minimization of the barrier function need not be exact. Unfortunately, barrier function algorithms suffer from numerical difficulties in the limit (e.g. ill conditioning of the Hessian matrix). Due to these difficulties these classical barrier methods became out-of-date in the 1970s.

The center method was introduced and studied by Huard [60]. As the logarithmic barrier function, this method also follows the central path to the optimum. Only the parameterization is different from the logarithmic barrier method. This method has attracted less attention in the (old) literature than the logarithmic barrier method for reasons which will become clear later on.

Dikin [31] proposed his affine scaling algorithm in 1967. This method does not have such a 'rich' history as barrier and center methods have. In fact this method remained unnoticed until recently, when it turned out that it is in fact a natural simplification of Karmarkar's projective method.

1.2 Intermezzo: Complexity issues

Since an IPM produces a sequence of interior solutions, a stopping rule that defines acceptable closeness to optimality is needed. Moreover, in complexity studies the order of magnitude of the number of computations (iterations, arithmetic operations, bit operations) needed to find an exact solution of the problem in the worst possible case is sought. A method is called polynomial if this number of computations is bounded by a polynomial in the so-called size of the problem.

Khachian [73] gave an explicit expression for the size of a linear problem, when he proposed the first polynomial algorithm for linear programming. Let us consider the primal linear programming problem

$$(\mathcal{LP}) \quad \begin{cases} \min \; c^T x \\ Ax = b \\ x \geq 0, \end{cases}$$

where A is an $m \times n$ matrix, and b and c are $m-$ and $n-$dimensional vectors respectively. It is assumed that the entries of A, b and c are integers. This is equivalent to assuming that the problem data are rational, since rational values can be rescaled to integers. The size of the problem (denoted by L) indicates the amount of information needed to represent an encoding of the problem. It is easy to verify that[2]

$$L = \Theta(mn + \lceil \log_2 |P| \rceil), \tag{1.1}$$

where P is the product of all nonzero coefficients in A, b and c (see [120]). A method is called polynomial if the number of operations (with finite precision) is bounded from above by a polynomial in L.

Khachian [73] showed, using Cramer's rule, that if x is a vertex of the feasible region, then no component of x can be in the interval $(0, 2^{-L})$. Moreover, if x is a vertex then $c^T x$ is either equal to the optimal value or the difference with the optimal value is at least 2^{-2L}. (See also Papadimitriou and Steiglitz [120].) In IPMs $c^T x - z^* \leq 2^{-2L}$ is used as a stopping rule, where z^* is the optimal value. From this point an $O(mn^2)$ rounding procedure (see [120], [83]) can be used to obtain a vertex with objective value less than or equal to $c^T x$, which must be optimal. We note that the optimal value need not be known for the stopping rule, since in most IPMs dual solutions are generated.

Note that to prove polynomiality we not only have to prove that the number of operations to obtain a 2^{-2L}-optimal solution is bounded by a polynomial in L, but also that all operations can be carried out in finite precision, more precisely, the number of bits is at most a polynomial in L. This issue is ignored in most papers on IPMs. In this book we measure the work in the number of arithmetic operations ($\times$, $/, +, -, \sqrt{}$) on real numbers rather than bit operations. (For an extensive discussion see [72], [127], [149].)

We will derive upper bounds for the number of iterations (or operations) to obtain an ϵ-optimal solution for several problems and algorithms. (A solution is called to be ϵ-optimal if its objective function value differs at most ϵ from the optimal value.) These upper bounds are all bounded by a polynomial in the problem dimensions and in $\ln \frac{\epsilon_0}{\epsilon}$, where ϵ_0 is the initial accuracy. For linear and quadratic programming this immediately implies polynomiality, i.e. the number of operations to obtain an exact optimal solution is bounded by a polynomial in L (take $\epsilon = 2^{-2L}$ and $\epsilon_0 = 2^{O(L)}$).

The definition of polynomiality given above is suitable for linear and quadratic programming, but for more complicated problems this definition cannot be used. Some authors ([119], [168] and [68]) have tried to define generalized polynomiality for more general optimization problems. In this book we will avoid this issue, we will only use the term polynomiality for linear and quadratic programming.

1.3 Classifying the IPMs

In this section we will briefly sketch the most important landmarks in the history of IPMs after Karmarkar's publication. A more detailed discussion will be given in the next chapters.

---

[2]The exact formula for L varies in the literature.

Soon after Karmarkar's publication, the original *projective potential reduction method* was studied and beautified by many researchers: simplification of the analysis, studies of limiting behavior, removing the initial assumptions, etc. From the practical point of view the early results, e.g. [141], were disappointing. Later more efficient algorithms were developed using better reformulations of the linear programming problem.

However, the practical merits of IPMs became clear when several researchers proposed and implemented the so–called *affine scaling method*. In each iteration of the projective potential reduction method a projective transformation is carried out to 'center' the current iterate. Instead of doing this (rather awkward) projective transformation, Vanderbei et al. [151] and Barnes [9] proposed to do a simple affine scaling to center the current iterate. It soon turned out that this natural simplification of Karmarkar's method was already proposed by Dikin [31] in 1967. Several implementations of the primal and dual affine scaling methods (e.g. [151], [1] and [107]) showed promising results.

Unfortunately, only global convergence under some conditions has been proved for these affine scaling methods. Due to the work of Megiddo and Shub [97], this method is believed not to be polynomial. The affine scaling method was generalized for quadratic programming by Ye and Tse [167] and to problems with convex objective function and linear constraints by Gonzaga and Carlos [56].

Soon after Karmarkar published his projective method, researchers became interested in the classical logarithmic barrier and the center method. As mentioned before, both methods are called *path–following methods*, since they follow the central path of the problem. We will now describe the main results obtained for this class of methods.

In 1986, Gill et al. [40] first derived a relationship between Karmarkar's method and the logarithmic barrier method for linear programming. They showed that the search directions used in Karmarkar's method and in the logarithmic barrier method were closely related. They also implemented the logarithmic barrier method, obtaining promising results. However, they did not prove polynomiality as Karmarkar did. This publication renewed the interest in barrier methods.

Gonzaga [47] was the first who proved polynomiality for a special version of the logarithmic barrier method for linear programming. In his method the barrier parameter is reduced by a factor $1 - \frac{0.005}{\sqrt{n}}$, which means that only very short steps are taken. Hence, although he proved that the total complexity is $O(n^3L)$, in practice this method is hopelessly slow. Ben Daya and Shetty [12] extended this method to quadratic programming. Monteiro and Adler [110], [111] developed a primal–dual (short–step) path–following method for linear and quadratic programming.

Later on, Roos and Vial [131] and Gonzaga [51] independently developed a natural and more practical version of the logarithmic barrier method for linear programming. In this method the barrier parameter may be reduced by an arbitrary factor between 0 and 1. This means that the iterates need not lie close to the central path and hence long steps can be taken. They also proved that this method has an $O(nL)$ iteration bound. In [30] we gave a simple analysis for the logarithmic barrier method and showed that the overall complexity can be reduced by a factor $\sqrt{n}$, using safeguarded line searches. In fact these recent developments show that the classical logarithmic barrier method can be made polynomial by using Newton's method for the internal minimization and using certain proximity criteria to stop this minimization.

In [7] we extended our analysis for the long–step analysis, given in [30], to quadratic programming. Moreover, in [29] we analyzed a natural variant of the classical log-arithmic barrier method for smooth convex programs. Nesterov and Nemirovsky [119] introduced the concept of self–concordance and using this notion, short–step barrier methods for smooth convex programming were studied.

Renegar [127] proved in 1987 that a very special version of the center method for lin-ear programming is polynomial. In fact, this was the first polynomial path–following method with an $O(\sqrt{n}L)$ iteration bound. In this method the relaxation parameter is very small $(\frac{1}{13\sqrt{n}})$, which means that very short steps are taken. Although this method for linear programming has an $O(\sqrt{n}L)$ iteration bound, which is a factor $\sqrt{n}$ better than the projective potential reduction method, it is very inefficient in practice. Later on Vaidya [149] reduced the total complexity bound to $O(n^3L)$, us-ing approximate solutions and rank–one updates. At this moment this complexity bound is still the best one obtained for any IPM, although there are more IPMs with this complexity bound. In [28] we developed a natural and more practical version of the center method for linear programming, in which the relaxation factor may be any number between 0 and 1, and line searches are allowed.

Sonnevend [135] suggested a method of centers for general convex programming. Jarre [67] presented first proofs of convergence for this method when applied to quadratically constrained convex quadratic programming problems. Later on, he described in [66] a special implementation of the method for a class of smooth convex programs. Again the relaxation parameter is very small, $\frac{1}{200(1+M)^2\sqrt{n}}$, where M is a curvature constant. He proved that the number of iterations to get an ϵ-optimal solution is polynomial in $|\log \epsilon|$, n and M. Similar results were obtained by Mehrotra and Sun [99] and Nesterov and Nemirovsky [119] in a even more general setting. Jarre also showed that extrapolation makes the method much more efficient. In [22] we proposed a more natural implementation of the classical center method for this class of smooth convex programming problems, in which the relaxation parameter for updating the lower bound may be any number between 0 and 1, and line searches are allowed. In that paper nice upper bounds for the total number of iterations have been obtained too.

After Renegar [127] published his short–step path–following method, and before an analysis for the long–step path–following method had been given, Ye [162] showed that the same complexity can be obtained without following the central path closely, and also allowing line searches. He defined a new so–called potential function, which is a 'surrogate' function used as a compass for finding the optimum. He proved that by doing projected gradient steps (after affine scaling) a constant reduction for the potential function value can be obtained. This method is called a *potential reduction method.* Kojima et al. [80] developed a primal–dual version of this method for linear complementarity problems. This primal–dual method with some additional tricks (see Mehrotra [98]) showed excellent practical behavior.

The number of papers on IPMs is enormous[3]. At this moment, eight years after Karmarkar's publication, it is safe to say that all IPMs can be categorized in the following 4 categories (in historical order) :

- Path–following methods;

- Affine scaling methods;

- Projective potential reduction methods;

- Affine potential reduction methods.

The differences between these classes are sometimes vague, and we will make clear in this book that all these methods rely on some common notions and concepts. One can also distinguish whether these algorithms are primal–only (dual–only) algorithms or primal–dual algorithms.

1.4 Scope of the book

The most important issue in this book is the complexity of an IPM. Our aim will be to derive upper bounds for the number of iterations (operations) needed by IPMs to obtain an ϵ–optimal solution for a certain problem. In our research we have focused our attention on path–following methods, more specifically on logarithmic barrier and center methods. The analysis of these methods can be nicely extended to convex programming.

We will give a complexity analysis for short–, medium– and long–step logarithmic barrier and center methods for linear, quadratic and convex programming. As indicated in the previous section, logarithmic barrier methods (short–, medium– and long–steps) for linear programming were already analyzed in the literature. We will simplify and unify the analysis, which also enables us to extend it to the quadratic and convex case. Moreover, for linear programming we will show how to reduce the computational load in each iteration. For the center method only a short–step variant was analyzed in the literature; we will also treat medium– and long–steps in a similar way as for the logarithmic barrier method and simplify the analysis for linear programming given in the literature.

The central ideas in the analysis for both the logarithmic barrier and the center method for linear, quadratic and convex programming are the same. First, the 'Hessian norm' of the Newton search direction is always used to measure the distance to the current reference point on the central path. Second, the quadratic

---

[3]There are about 2000 papers which deal with IPMs! (See also Kranich's [85] bibliography of IPM papers.)

convergence result in the vicinity of the central path enables us to prove many other basic properties, which are needed for the complexity analysis.

In Chapter 2 we will deal with the logarithmic barrier method for linear [30], quadratic [7] and smooth convex programming [29], [26]. We will analyze long--, medium-- and short--step variants. We note that our analysis for smooth convex programming in [29] is somewhat different from our analysis for linear [30] and quadratic programming [7]. The analysis for smooth convex programming given in this book has the same structure as the analysis for linear and quadratic programming. This unified treatment is based on the notion of self--concordance.

In Chapter 3 we will deal with the center method for linear [22] and smooth convex programming [28]. Again, contrary to our analysis in [28], we will treat both cases in a similar setting. Moreover, we will show the similarities and differences with the logarithmic barrier method. Due to the relationship between the logarithmic barrier and center method, many results follow more or less directly from Chapter 2. Again, we will analyze three variants: long--, medium-- and short--step variants.

In Chapter 4 we will show that using approximate solutions, rank--one updates and safeguarded line searches, the overall complexity for linear programming can be reduced by a factor $\sqrt{n}$ [30]. Moreover, we will show that the amount of work in each iteration can be reduced by using some kind of column generation and deletion technique [25], [27].

In Chapter 5 we will give a (short) description of other IPMs (affine scaling, projective and affine potential reduction methods), and indicate some open problems. We will show that all these methods rely on the same notions: they all use the central path explicitly or implicitly and the search directions used are all linear combinations of two characteristic vectors [21]. Moreover, we will briefly discuss some computational results for IPMs.

Finally, in Chapter 6 we end up with a short summary, some conclusions and recommendations for future research. The Appendices contain self--concordance proofs for some classes of problems [19], and some general technical lemmas which are used in this book. For the notation used in this book we refer the reader to the Glossary.

Chapter 2

The logarithmic barrier method

In this chapter we analyze the complexity of special variants of the logarithmic barrier method for some classes of problems. We will deal with so–called long–, medium– and short–step methods. First, we will give a general framework for the logarithmic barrier method. Then we look at special cases: linear, convex quadratic and smooth convex programming problems respectively.

The quadratic convergence result in the vicinity of the central path is used to derive upper bounds for the differences in barrier function value and objective value between a point on the central path and a point in its vicinity. The linear programming problem is stated in inequality form. Using the null–space formulation, we are able to generalize the results to convex quadratic programming. For a change, we will deal with the quadratic problem in equality form. The analysis for the convex programming problem is more difficult than the analysis for linear and quadratic programming, but the structures are similar. The self–concordance property, introduced by Nesterov and Nemirovsky [119], plays a key role in this analysis.

2.1 General framework

We consider the primal formulation of the convex programming problem:

$$(\mathcal{CP}) \quad \begin{cases} \max\ f_0(y) \\ f_i(y) \leq 0, \quad i = 1, \cdots, n \\ y \in \mathbb{R}^m. \end{cases}$$

The feasible region is denoted by $\mathcal{F}$, and the interior of this region by $\mathcal{F}^0$. We will assume that:

Assumption 1: the functions $-f_0(y)$ and $f_i(y)$, $1 \leq i \leq n$, are convex functions with continuous first and second order derivatives in $\mathcal{F}^0$;

Assumption 2: $\mathcal{F}^0$ is nonempty;

Assumption 3: $\mathcal{F}^0$ is bounded.

9

Wolfe's [152] formulation of the dual problem associated with this primal problem
is

$$(\mathcal{CD}) \quad \begin{cases} \min \; f_0(y) - \sum_{i=1}^{n} x_i f_i(y) \\ \sum_{i=1}^{n} x_i \nabla f_i(y) = \nabla f_0(y) \\ x_i \geq 0. \end{cases}$$

At the risk of labouring the obvious, we want to point out that $(\mathcal{CD})$ is not necessarily
convex. The following theorem is a well–known result (see [152]).

Theorem 2.1 *If y is a feasible solution of $(\mathcal{CP})$ and $(\bar{y}, x)$ is a feasible solution of
the dual problem $(\mathcal{CD})$, then*

$$f_0(y) \leq f_0(\bar{y}) - \sum_{i=1}^{n} x_i f_i(\bar{y}).$$

Proof: The proof can easily be established with the properties of convex functions.
For a convex function $f(y)$ we have

$$f(\bar{y}) \leq f(y) + (\bar{y} - y)^T \nabla f(\bar{y}).$$

Since $-f_0(y)$ and $f_i(y)$, $1 \leq i \leq n$, are convex functions we find

$$\begin{aligned} f_0(\bar{y}) - \sum_{i=1}^{n} x_i f_i(\bar{y}) \; &\geq \; f_0(y) + (\bar{y} - y)^T \nabla f_0(\bar{y}) \\ &\quad - \sum_{i=1}^{n} x_i f_i(y) - \sum_{i=1}^{n} x_i (\bar{y} - y)^T \nabla f_i(\bar{y}) \\ &= \; f_0(y) - \sum_{i=1}^{n} x_i f_i(y) \\ &\geq \; f_0(y). \end{aligned}$$

The equality and the last inequality follows because $(\bar{y}, x)$ is a feasible solution of
the dual problem $(\mathcal{CD})$ and y is a feasible solution of the primal problem $(\mathcal{CP})$. $\square$

Note that Assumption 2 is in fact a Slater condition. Due to the assumptions, $(\mathcal{CD})$
has an optimal solution, and the extremal values of $(\mathcal{CP})$ and $(\mathcal{CD})$ are equal.

The logarithmic barrier function associated with $(\mathcal{CP})$ is

$$\phi_{\text{B}}(y, \mu) = -\frac{f_0(y)}{\mu} - \sum_{i=1}^{n} \ln(-f_i(y)), \qquad (2.1)$$

where $\mu > 0$ is the barrier parameter. Because of the singularity of the logarithm at
zero, this barrier function will prevent the iterates from going outside the feasible
region. Therefore the logarithmic barrier function method is called an interior point
method.

For the description of the logarithmic barrier method we need the first and second order derivatives of $\phi_B(y,\mu)$:

$$g(y,\mu) := \nabla\phi_B(y,\mu) = -\frac{\nabla f_0(y)}{\mu} + \sum_{i=1}^{n}\frac{\nabla f_i(y)}{-f_i(y)} \tag{2.2}$$

and

$$\begin{aligned}H(y,\mu) &:= \nabla^2\phi_B(y,\mu)\\ &= -\frac{\nabla^2 f_0(y)}{\mu} + \sum_{i=1}^{n}\left[\frac{\nabla^2 f_i(y)}{-f_i(y)} + \frac{\nabla f_i(y)\nabla f_i(y)^T}{f_i(y)^2}\right].\end{aligned} \tag{2.3}$$

If no confusion is possible we will write, for shortness sake, g and H instead of $g(y,\mu)$ and $H(y,\mu)$.

It is obvious that $H(y,\mu)$ is positive semidefinite, but the Assumptions 1–3 are not sufficient to guarantee that $H(y,\mu)$ is positive definite. The following example, given by Jarre [66], illustrates this. Let $m = n = 1$, $f_0(y) = 0$ and

$$f_1(y) = \begin{cases} y^4 - 1, & y \le 0 \\ -1, & 0 < y < 1 \\ (y-1)^4 - 1, & y \ge 1. \end{cases} \tag{2.4}$$

It is easy to verify that in this example Assumptions 1–3 are satisfied, but for $0 < y < 1$ we have $H(y,\mu) = 0$.

However, for the classes of problems we will consider in the next sections (linear programming, convex quadratic programming and convex programming satisfying the self–concordance condition) we will show that $H(y,\mu)$ is positive definite on its domain $\mathcal{F}^0$, i.e. $\phi_B(y,\mu)$ is strictly convex. So, in the remainder of this section we assume that $H(y,\mu)$ is positive definite.

We will use the measure $\|.\|_H$ to measure distances between points. The definition of this measure is as follows:
$$\|z\|_H = \sqrt{z^T H z}.$$

Because H is positive definite, $\|.\|_H$ defines a norm. We emphasize that this norm depends on y, since $H = H(y,\mu)$ depends on y.

Since $\phi_B(y,\mu)$ is strictly convex on its bounded domain $\mathcal{F}^0$ and takes infinite values on the boundary of $\mathcal{F}$, this function achieves the minimal value in its domain (for fixed μ) at a unique point, which is denoted by $y(\mu)$, and called the μ–center. The necessary and sufficient Karush–Kuhn–Tucker conditions for $y(\mu)$ are:

$$\begin{cases} f_i(y) \le 0, & 1 \le i \le n \\ \sum_{i=1}^{n} x_i \nabla f_i(y) = \nabla f_0(y), & x \ge 0 \\ -f_i(y)x_i = \mu, & 1 \le i \le n. \end{cases} \tag{2.5}$$

Definition of the central path: The primal (dual) central path (or trajectory) is defined as the set of centers $y(\mu)$ $(x(\mu), y(\mu))$, where μ runs from ∞ to 0.

Note that not only $y(\mu)$ is primal feasible, but also $(x(\mu), y(\mu))$ is dual feasible. Moreover, the duality gap in $(x(\mu), y(\mu), s(\mu))$ satisfies

$$f_0(y(\mu)) - \sum_{i=1}^n x_i(\mu) f_i(y(\mu)) - f_0(y(\mu)) = - \sum_{i=1}^n x_i(\mu) f_i(y(\mu)) = n\mu. \qquad (2.6)$$

It is well–known that $x(\mu)$ and $y(\mu)$ are continuously differentiable[1] (see [35]). Hence, it holds that $y(\mu)$ and $x(\mu)$ will converge to optimal solutions of $(\mathcal{CP})$ and $(\mathcal{CD})$, if $\mu \downarrow 0$. This means that the central trajectory ends in an optimal solution of the problem. Moreover, note that if the feasible region is bounded then the central path starts in the unique point where $\sum_{i=1}^n \ln(-f_i(y))$ is maximized, since

$$\lim_{\mu \to \infty} \phi_B(y, \mu) = - \sum_{i=1}^n \ln(-f_i(y)).$$

This point is called the *analytic center* of the feasible region.

The central path was studied by Fiacco and McCormick [35] and McLinden [94] for the general convex case, and by Megiddo [96], Bayer and Lagarias [11] and Sonnevend [135] for the linear case. In Section 2.2 we will give some examples of central paths for some special problems.

The logarithmic barrier function method was introduced by Frisch [37], [38] and further developed by Fiacco and McCormick [35]. In this method the original constrained problem $(\mathcal{CP})$ is replaced by a sequence of unconstrained minimization problems, i.e. minimizing $\phi_B(y, \mu)$ successively for a sequence of positive decreasing values of the barrier parameter μ. Loosely speaking, this method follows the central path approximately to an optimal point.

There are some important elements in the design of such a method:

1. the method used to (approximately) minimize $\phi_B(y, \mu)$;

2. the criterion to terminate this approximate minimization;

3. the updating scheme for the barrier parameter μ.

Fiacco and McCormick [35] recommended to do line searches along Newton directions, but did not treat the problem when to terminate the approximate minimization; for fixed μ the logarithmic barrier function is minimized exactly. The updating scheme they suggest is simply to reduce μ by a constant smaller than one. This sequential unconstrained minimization technique (SUMT) was implemented in the SUMT–3 and SUMT–4 codes, described in Mylander et al. [117]. Anstreicher [4] obtained, using the analysis in [7] (see also Section 2.4), the surprising result that this algorithm (exactly as implemented in these codes) solves linear and quadratic

---

[1]This can be proved by using the implicit function theorem.

programs in $O(\sqrt{n}L \ln L)$ iterations, with proper initialization and choice of parameters.

The algorithm we will propose and analyze below also does a line search along the Newton direction. The Newton direction is defined as $p = -H^{-1}g$. In the analysis we will see that taking a suitable steplength is already sufficient for the complexity analysis. The criterion to terminate the approximate minimization of $\phi_{\text{B}}(y, \mu)$ is $\|p\|_H \leq \tau < 1$. Note that $\|p\|_H = 0$ if and only if $y = y(\mu)$. This proximity measure, which is in fact the Hessian norm of the search direction, will appear to be very appropriate. For linear and quadratic programming we will develop a nice characterization for this measure, which enables us to prove quadratic convergence easily. We will analyze several updating schemes for the barrier parameter.

We now describe the algorithm for finding an ϵ-optimal solution.

Logarithmic Barrier Algorithm

Input:

ϵ is the accuracy parameter;

τ is the proximity parameter;

θ is the reduction parameter, $0 < \theta < 1$;

μ_0 is the initial barrier value;

y^0 is a given interior feasible point such that $\|p(y^0, \mu_0)\|_{H(y^0, \mu_0)} \leq \tau$;

begin
    $y := y_0;\ \mu := \mu_0;$
    while $\mu > \frac{\epsilon}{4n}$ **do**
    begin (outer step)
        $\mu := (1 - \theta)\mu;$
        while $\|p\|_H \geq \tau$ **do**
        begin (inner step)
            $\tilde{\alpha} := \arg\min_{\alpha > 0} \{\phi_{\text{B}}(y + \alpha p, \mu) : y + \alpha p \in \mathcal{F}^0\}$
            $y := y + \tilde{\alpha} p$
        end (inner step)
    end (outer step)
end.

As indicated above, the line search need not be exact; a suitable steplength will be sufficient for the complexity analysis. For finding an initial point that satisfies the input assumptions of the algorithm we refer the reader to Renegar [127], Monteiro and Adler [110] and Güler et al. [57] for linear programming, Ye [158] for quadratic programming, and Jarre [66] and Mehrotra and Sun [99] for convex programming. Later on the 'centering assumption' will be alleviated.

We introduce some terminology. We call the algorithm a

- **long–step** algorithm if θ is a constant ($0 < \theta < 1$), independent of n and ϵ;

- **medium–step** algorithm if $\theta = \frac{\nu}{\sqrt{n}}$, where $\nu > 0$ is an arbitrary constant, possibly large, and independent of n and ϵ;

- **short–step** algorithm if $\theta = \frac{\nu}{\sqrt{n}}$, and ν is so small (e.g. $\frac{1}{9}$), that after a reduction of μ one unit Newton step is sufficient to reach the vicinity of the new μ–center.

Figure 2.1 shows some iterations for the long– or medium–step algorithm for a linear programming example. Figure 2.2 shows some short–step iterations.

In this chapter we will analyze such long–, medium– and short–step algorithms. Note that the long–step algorithm is the most promising for practical use. However, we will obtain the remarkable result that the theoretical complexity of the long–step algorithm is worse than the complexity of short– and medium–step algorithms.

2.2   Central paths for some examples

In this section we will illustrate the theory of central paths by calculating these paths for some explicit examples.

Example 1. Let us consider the following simple linear programming problem ($m = 2$ and $n = 4$):

$$\begin{cases} \max 2y_1 + y_2 \\ y_1 \leq 1 \\ -y_1 \leq 1 \\ y_2 \leq 1 \\ -y_2 \leq 1. \end{cases}$$

The logarithmic barrier function for this problem is:

$$\phi_{\text{B}}(y, \mu) = -\frac{2y_1 + y_2}{\mu} - \ln(1 - y_1)$$
$$- \ln(1 - y_2) - \ln(1 + y_1) - \ln(1 + y_2).$$

This function is minimized by setting the derivatives (with respect to y_1 and y_2) equal to zero. Solving these equations we obtain

$$y_1(\mu) = \frac{1}{2}\sqrt{\mu^2 + 4} - \frac{1}{2}\mu$$

and

$$y_2(\mu) = \sqrt{\mu^2 + 1} - \mu.$$

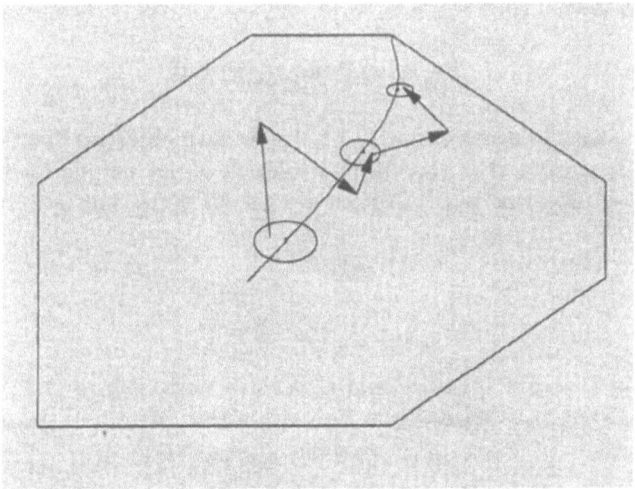

Figure 2.1: Some iterations of the long– or medium–step method.

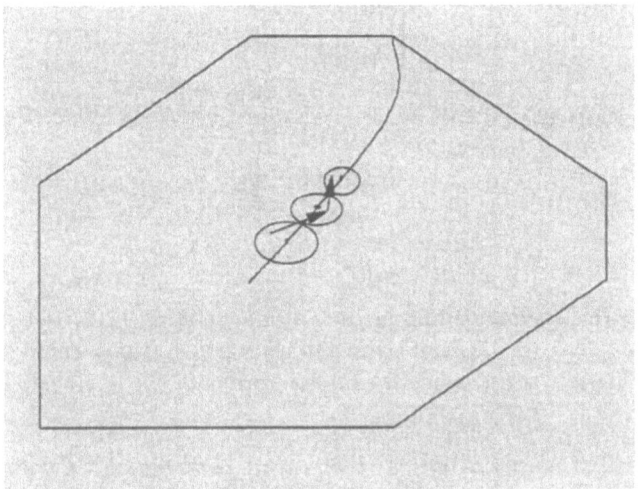

Figure 2.2: Some iterations of the short–step method.

Note that this central path really ends in the optimal solution $(1,1)$ since

$$\lim_{\mu \to 0} y_1(\mu) = \lim_{\mu \to 0} y_2(\mu) = 1.$$

Moreover, this central path starts in $(0,0)$, the analytic center of the feasible region, since

$$\lim_{\mu \to \infty} y_1(\mu) = \lim_{\mu \to \infty} y_2(\mu) = 0.$$

Example 2. Consider again Example 1 but now with objective function y_2 instead of $2y_1 + y_2$. It is clear that the optimal solution is not unique for this example. Doing the same algebraic manipulations we get for $y_1(\mu)$ and $y_2(\mu)$ the following expressions:

$$y_1(\mu) = 0$$

and

$$y_2(\mu) = \sqrt{\mu^2 + 1} - \mu.$$

By letting μ go to zero it is clearly that the central path ends in $(0,1)$, which is one of the optimal solutions. Moreover $(0,1)$ is the analytic center of the optimal facet, i.e. it maximizes $-\sum_{i=1}^{n} \ln(-f_i(y))$ over the optimal facet[2].

Example 3a. Let us now consider the problem[3]

$$\begin{cases} \max\ 2y_1 + y_2 \\ y_1 \leq 1 \\ y_2 \leq 1 \end{cases}$$

By doing the same calculations as for Examples 1 and 2 we obtain that

$$y_1(\mu) = 1 - \frac{1}{2}\mu$$

and

$$y_2(\mu) = 1 - \mu.$$

The central path is a line ending in the optimal solution $(1,1)$ (let μ go to zero). Moreover, note that this central path has no starting point: there is no analytic center of the feasible region since the feasible region is not bounded.

Example 3b. The observation made in Example 3a (the central path is a line) is true also in a more general setting: if the feasible region of a linear program is a cone then the central path is a line, ending in the top of the cone. We will show this now. Suppose that the constraints of the linear problem describe a cone in

---

[2]It is a general property that the central path ends in the analytic center of the set of optimal solutions (see e.g. Megiddo [96]).

[3]The feasible region of this problem is not bounded, however for linear programming the boundedness assumption can be alleviated to bounded level sets, which holds in this case. See also Section 2.3.

$\mathbb{R}^m$. Moreover, let the objective function be such that the top of the cone is the optimal solution. Let the constraint matrix be denoted by A^T, which is a square and invertible $m \times m$ matrix, and let b, an m–dimensional vector, be the objective vector. So the problem is as follows

$$\begin{cases} \max \ b^T y \\ A^T y \le c. \end{cases}$$

Note that $t = A^{-T}c$ is the top of the cone. The logarithmic barrier function for this problem is

$$\phi_B(y, \mu) = -\frac{b^T y}{\mu} - \sum_{i=1}^{m} \ln(c_i - a_i^T y),$$

where a_i is the i-th row of A^T. It can easily be verified that (2.5) reduces to[4]

$$\begin{cases} A^T y + s = c, & s \ge 0 \\ Ax = b, & x \ge 0 \\ Sx = \mu e. \end{cases} \tag{2.7}$$

From the second equation of (2.7) we get

$$x(\mu) = A^{-1}b$$

and from the third

$$s(\mu) = \mu X^{-1} e.$$

Substituting this into the first equation we obtain

$$\begin{aligned} y(\mu) &= A^{-T}(c - s) \\ &= t - \mu v, \end{aligned}$$

where

$$v_i = \frac{1}{(A^{-1}b)_i}.$$

Consequently, the central path is a line ending in the optimal point, namely the top of the cone!

Example 4. Let us now consider the following simple convex quadratic programming problem ($m = 2$ and $n = 1$):

$$\begin{cases} \max -y_1 - y_2 \\ y_1^2 + 2y_2^2 \le 1 \end{cases}$$

---

[4]See also Section 2.3.

The logarithmic barrier function for this problem is:

$$\phi_B(y, \mu) = \frac{y_1 + y_2}{\mu} - \ln(1 - y_1^2 - 2y_2^2).$$

This function is minimized by setting the derivatives (with respect to y_1 and y_2) equal to zero. Solving these equations we obtain

$$y_1(\mu) = \frac{2}{3}\mu - \frac{1}{3}\sqrt{4\mu^2 + 6}$$

and

$$y_2(\mu) = \frac{1}{3}\mu - \frac{1}{6}\sqrt{4\mu^2 + 6}.$$

Note that this central path really ends in the optimal solution $(-\frac{2}{\sqrt{6}}, -\frac{1}{\sqrt{6}})$ since

$$\lim_{\mu \to 0} y_1(\mu) = -\frac{2}{\sqrt{6}}$$

and

$$\lim_{\mu \to 0} y_2(\mu) = -\frac{1}{\sqrt{6}}.$$

Moreover, this central path starts in $(0,0)$, the analytic center of the feasible region, since

$$\lim_{\mu \to \infty} y_1(\mu) = \lim_{\mu \to \infty} y_2(\mu) = 0.$$

Example 5. For the above three examples we were able to give an explicit expression for the central path. This is, of course, in general not possible. The following linear programming problem, with $m = 2$ and $n = 7$ is an example of this:

$$\begin{cases} \max y_1 \\ y_1 \leq 1 \\ -y_1 \leq 1 \\ y_2 \leq 1 \\ -y_2 \leq 1 \\ -y_1 + y_2 \leq 1.25 \\ y_1 + y_2 \leq 1.25 \\ y_1 - y_2 \leq 1.25 \end{cases}$$

It can easily be verified for this example that it is not possible to give an explicit expression for $y_1(\mu)$ and $y_2(\mu)$. Figure 2.3 shows some central paths for this problem with different objective functions. The objective functions are

$$(\cos \frac{i}{6}\pi)y_1 + (\sin \frac{i}{6}\pi)y_2, \quad i = 0, 1, \cdots, 11.$$

Note that for some objective functions (e.g. the objective function y_1) a whole facet is optimal. In these cases, as we have seen above, the central path ends in the analytic center of the optimal facet.

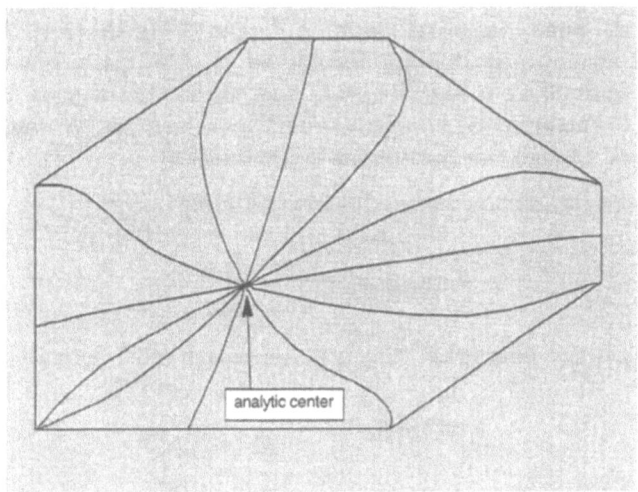

Figure 2.3: Central paths for some LP problems with the same feasible region and different objective vectors.

2.3 Linear programming

2.3.1 Properties on and near the central path

In this section we will deal with the linear programming problem in inequality form[5]:

$$(\mathcal{LD}) \quad \begin{cases} \max\ b^T y \\ A^T y \le c. \end{cases}$$

Here A is an $m \times n$ matrix, b and c are $m-$ and $n-$ dimensional vectors respectively; the $m-$dimensional vector y is the variable in which the maximization is carried out. The slack variable s is defined as $s := c - A^T y$.

The primal problem for $(\mathcal{LD})$ is:

$$(\mathcal{LP}) \quad \begin{cases} \min\ c^T x \\ Ax = b \\ x \ge 0. \end{cases}$$

The analysis which we will give in this section can also be applied to $(\mathcal{LP})$ (see [30]). We will not do this here, instead in the next section we will treat the convex quadratic programming problem in equality form.

---

[5]Contrary to the formulation of $(\mathcal{CP})$, we call this problem the dual problem, to be consistent with the LP literature.

Without loss of generality we assume that all coefficients are integral. We shall denote by L the binary length of the input data of $(\mathcal{LP})$; see (1.1). We also make the standard assumptions that the feasible set of $(\mathcal{LD})$ has a nonempty interior and that the optimal set is bounded. Note that the last assumption is weaker than Assumption 3. Furthermore, in order to simplify the analysis, we shall assume that A has full rank, though this assumption is not essential.

We consider the logarithmic barrier function for $(\mathcal{LD})$

$$\phi_{\mathrm{B}}(y,\mu) := -\frac{b^T y}{\mu} - \sum_{i=1}^{n} \ln s_i, \tag{2.8}$$

where μ is a positive parameter. The first and second order derivatives of ϕ_{B} are

$$g := \nabla \phi_{\mathrm{B}}(y,\mu) = -\frac{b}{\mu} + AS^{-1}e,$$

$$H := \nabla^2 \phi_{\mathrm{B}}(y,\mu) = AS^{-2}A^T.$$

Under the assumptions made above, ϕ_{B} achieves a minimum value over $\mathcal{F}$ at a unique point; see e.g. [96]. The necessary and sufficient first order optimality conditions for this point are (see (2.5)):

$$\begin{cases} A^T y + s = c, & s \geq 0 \\ Ax = b, & x \geq 0 \\ Sx = \mu e. \end{cases} \tag{2.9}$$

It is easy to verify that the necessary and sufficient first order optimality conditions for the minimum of the primal logarithmic barrier function, given by

$$\phi_{\mathrm{B}}^p(x,\mu) = \frac{c^T x}{\mu} - \sum_{i=1}^{n} \ln x_i, \tag{2.10}$$

are also given by (2.9).

We denote by $(x(\mu), y(\mu), s(\mu))$, the unique solution of (2.9). The duality gap in these points satisfies (see (2.6))

$$x(\mu)^T s(\mu) = n\mu. \tag{2.11}$$

It is well–known that $x(\mu)$ and $y(\mu)$ are continuously differentiable (see e.g. [35] and [96]). Hence, if $\mu \downarrow 0$ then $x(\mu)$ and $y(\mu)$ will converge to optimal primal and dual solutions respectively.

The following lemma states that the primal objective decreases along the primal path and the dual objective increases along the dual path. These results follow from Fiacco and McCormick [35].

Lemma 2.1 *If μ decreases then the objective $c^T x(\mu)$ of the primal problem $(\mathcal{LP})$ is monotonically decreasing and the objective $b^T y(\mu)$ of the dual problem $(\mathcal{LD})$ is monotonically increasing.*

Proof: Suppose $\bar{\mu} < \mu$. Since $y(\mu)$ minimizes $\phi_B(y, \mu)$ and $y(\bar{\mu})$ minimizes $\phi_B(y, \bar{\mu})$ we obviously have

$$\phi_B(y(\mu), \mu) \leq \phi_B(y(\bar{\mu}), \mu)$$

and

$$\phi_B(y(\bar{\mu}), \bar{\mu}) \leq \phi_B(y(\mu), \bar{\mu}),$$

which are equivalent to, respectively,

$$-\frac{b^T y(\mu)}{\mu} - \sum_{i=1}^{n} \ln s_i(\mu) \leq -\frac{b^T y(\bar{\mu})}{\mu} - \sum_{i=1}^{n} \ln s_i(\bar{\mu}),$$

and

$$-\frac{b^T y(\bar{\mu})}{\bar{\mu}} - \sum_{i=1}^{n} \ln s_i(\bar{\mu}) \leq -\frac{b^T y(\mu)}{\bar{\mu}} - \sum_{i=1}^{n} \ln s_i(\mu).$$

Adding the two inequalities gives

$$-\frac{b^T y(\mu)}{\mu} - \frac{b^T y(\bar{\mu})}{\bar{\mu}} \leq -\frac{b^T y(\bar{\mu})}{\mu} - \frac{b^T y(\mu)}{\bar{\mu}},$$

or equivalently

$$\left(\frac{1}{\bar{\mu}} - \frac{1}{\mu}\right)(b^T y(\mu) - b^T y(\bar{\mu})) \leq 0.$$

Since $\bar{\mu} < \mu$ the second part of the lemma follows. The first part of the lemma can be proved in the same way, since $x(\mu)$ is the minimizing point for the primal logarithmic barrier function $\phi_B^p(x, \mu)$. $\qquad\square$

Roos and Vial [133] introduced the following measure for the distance of an interior feasible point y to $y(\mu)$:

$$\delta(y, \mu) := \min_x \left\{ \left\| \frac{Sx}{\mu} - e \right\| : Ax = b \right\}. \qquad (2.12)$$

The unique solution of the minimization problem in the definition of $\delta(y, \mu)$ is denoted by $x(y, \mu)$. So, we can also write

$$\delta(y, \mu) = \left\| \frac{Sx(y, \mu)}{\mu} - e \right\|. \qquad (2.13)$$

It can easily be verified that

$$y = y(\mu) \iff \delta(y, \mu) = 0,$$

and moreover,

$$\delta(y, \mu) = 0 \implies x(y, \mu) = x(\mu).$$

The Newton direction $p(y, \mu)$ in y with respect to $\phi_B(y, \mu)$ is given by

$$p(y, \mu) = -H^{-1}g = (AS^{-2}A^T)^{-1}\left(\frac{b}{\mu} - AS^{-1}e\right). \qquad (2.14)$$

In the sequel of this section we will also write p instead of $p(y, \mu)$. The next lemma states that there is a close relationship between the δ–measure and the Hessian norm of the Newton direction $p(y, \mu)$.

Lemma 2.2 *For given y and μ we have*

$$\delta(y, \mu) = \|p(y, \mu)\|_H = \|S^{-1}A^T p(y, \mu)\|.$$

Proof: The last equality $\|p(y, \mu)\|_H = \|S^{-1}A^T p(y, \mu)\|$ holds by definition. To prove the first equality we derive from (2.12) an explicit expression for $x(y, \mu)$:

$$
\begin{aligned}
x(y, \mu) &= \mu S^{-1}e + \mu S^{-2}A^T(AS^{-2}A^T)^{-1}(\frac{b}{\mu} - AS^{-1}e) \\
&= \mu S^{-1}e + \mu S^{-2}A^T p, \qquad (2.15)
\end{aligned}
$$

where the last equality follows from (2.14). This means that

$$\frac{Sx(y, \mu)}{\mu} - e = S^{-1}A^T p, \qquad (2.16)$$

which proves the first equality of the lemma.   □

Note that

$$p_s := -A^T p \qquad (2.17)$$

is the search direction in the s–space. Consequently, from Lemma 2.2 it follows that

$$\delta(y, \mu) = \|S^{-1}p_s\|, \qquad (2.18)$$

i.e. the δ–measure is in fact the length of the scaled search direction in the s–space.

Now we will prove some fundamental lemmas for nearly centered points. The following quadratic convergence lemma is due to Roos and Vial [133] and is illustrated in Figure 2.4.

Lemma 2.3 *If $\delta(y, \mu) < 1$, then $y^+ = y + p$ is a strictly feasible point for (LD). Moreover,*

$$\delta(y^+, \mu) \le \delta(y, \mu)^2.$$

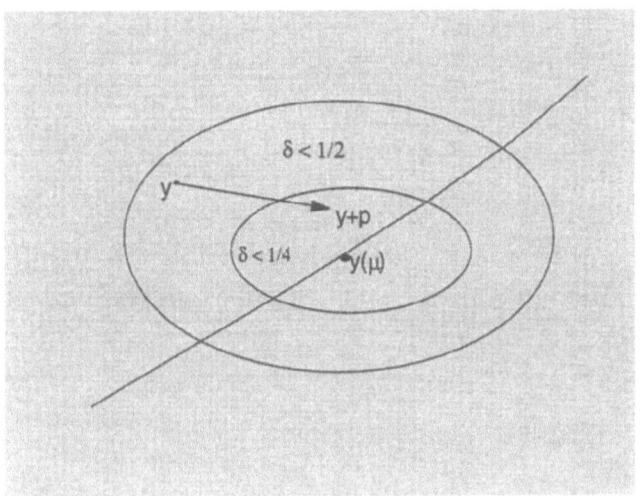

Figure 2.4: Quadratic convergence in the vicinity of a μ-center.

Proof: The proof is a simplified version of the proof in Roos and Vial [133]. Since $s^+ = s + p_s$, it follows from (2.17) that

$$A^T y^+ + s^+ = A^T(y + p) + s - A^T p = c.$$

Moreover, $s^+ = s + p_s = S(e + S^{-1} p_s) > 0$, since $\|S^{-1} p_s\| < 1$. This proves the first part of the lemma.

The definition of $x(y^+, \mu)$ implies the following:

$$\delta(y^+, \mu) = \left\| \frac{S^+ x(y^+, \mu)}{\mu} - e \right\| \leq \left\| \frac{S^+ x(y, \mu)}{\mu} - e \right\|.$$

Now using (2.15) and (2.18) and the fact that $S^+ = S + P_s$, we find

$$
\begin{aligned}
\delta(y^+, \mu) &\leq \left\| \frac{(S + P_s)(\mu S^{-1} e - \mu S^{-2} p_s)}{\mu} - e \right\| \\
&= \| P_s S^{-2} p_s \| \\
&\leq \| S^{-1} p_s \|^2 \\
&= \delta(y, \mu)^2.
\end{aligned}
$$

□

The following lemma gives an upper bound for the difference in barrier function value in a nearly centered point y and the exact μ-center $y(\mu)$.

Lemma 2.4 *If $\delta := \delta(y, \mu) < 1$ then*

$$\phi_{\text{B}}(y, \mu) - \phi_{\text{B}}(y(\mu), \mu) \leq \frac{\delta^2}{1 - \delta^2}.$$

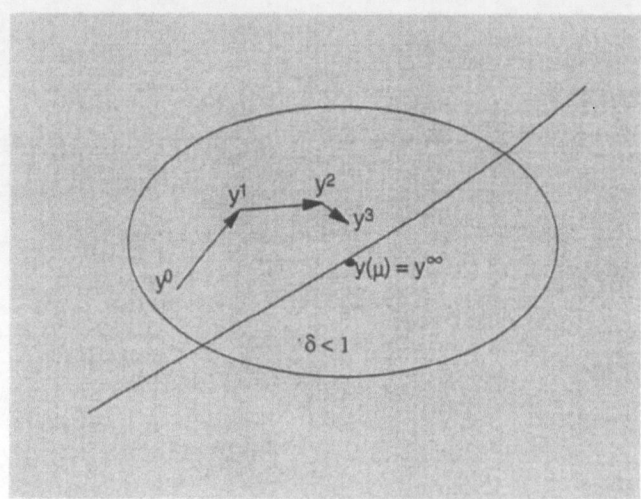

Figure 2.5: A sequence of Newton steps converging to the μ–center.

Proof: The barrier function ϕ_B is convex in y, whence

$$\phi_B(y + p, \mu) \geq \phi_B(y, \mu) + p^T g.$$

Now using $p = -H^{-1}g$, we have

$$p^T g = -p^T Hp = -\delta^2, \tag{2.19}$$

where the last equality follows from Lemma 2.2. Substitution gives

$$\phi_B(y, \mu) - \phi_B(y + p, \mu) \leq \delta^2. \tag{2.20}$$

Now let $y^0 := y$ and let $y^1, y^2, \ldots$ denote the sequence of points obtained by repeating Newton steps, starting at y^0 (see Figure 2.5). By Lemma 2.3 we have

$$\delta(y^i, \mu) \leq \delta(y^0, \mu)^{2^i} = \delta^{2^i}. \tag{2.21}$$

This means that $\delta(y^i, \mu) \to 0$ if $i \to \infty$. Since all y^i lie in a bounded region[6] and $\delta(y, \mu) = 0$ if and only if $y = y(\mu)$, we have that all limit points of the sequence are $y(\mu)$. Hence, the sequence of Newton iterates converges to $y(\mu)$. So, using (2.20) and (2.21), we may write

$$\phi_B(y, \mu) - \phi_B(y(\mu), \mu) \;=\; \sum_{i=0}^{\infty} \Big(\phi_B(y^i, \mu) - \phi_B(y^{i+1}, \mu) \Big)$$

---

[6]Lemma 2.7 tells us that all y^i lie in a certain level set, and since we assumed that the optimal set is bounded all level sets are bounded.

$$\leq \sum_{i=0}^{\infty} \delta(y^i, \mu)^2$$

$$\leq \sum_{i=0}^{\infty} \delta^{2^{i+1}}$$

$$\leq \frac{\delta^2}{1 - \delta^2}.$$

□

The following lemma gives an upper bound for the difference in objective function value in a nearly centered point y and $y(\mu)$.

Lemma 2.5 *If* $\delta := \delta(y, \mu) < 1$, *then*

$$|b^T y - b^T y(\mu)| \leq \frac{\delta(1 + \delta)}{1 - \delta} \mu \sqrt{n}.$$

Proof: From (2.19) we have $p^T g = -\delta^2$. On the other hand

$$
\begin{aligned}
-p^T g &= p^T \left(\frac{b}{\mu} - AS^{-1}e \right) \\
&= \frac{b^T p}{\mu} - e^T S^{-1} A^T p.
\end{aligned}
$$

So we have

$$\frac{b^T p}{\mu} = \delta^2 + e^T S^{-1} A^T p.$$

Using the Cauchy–Schwartz inequality, we obtain

$$|e^T S^{-1} A^T p| \leq \|S^{-1} A^T p\| \, \|e\| = \delta \sqrt{n},$$

where the last equality follows from Lemma 2.2. From this we deduce that

$$
\begin{aligned}
|b^T p| &\leq \mu(\delta^2 + \delta \sqrt{n}) \\
&= \delta(1 + \frac{\delta}{\sqrt{n}}) \mu \sqrt{n} \\
&\leq \delta(1 + \delta) \mu \sqrt{n}.
\end{aligned}
$$

Again, let $y^0 := y$ and let $y^1, y^2, \cdots$ denote the sequence of points obtained by repeating Newton steps, starting at y^0 and, as proved in Lemma 2.4 converging to $y(\mu)$ (see Figure 2.5). Then we have

$$
\begin{aligned}
|b^T y(\mu) - b^T y| &= \left| \sum_{i=0}^{\infty} \left(b^T y^{i+1} - b^T y^i \right) \right| \\
&\leq \sum_{i=0}^{\infty} \left| b^T p(y^i, \mu) \right|
\end{aligned}
$$

$$\leq \sum_{i=0}^{\infty} \delta(y^i, \mu)(1 + \delta(y^i, \mu))\mu\sqrt{n}$$

$$\leq \sum_{i=0}^{\infty} \delta^{2^i}(1 + \delta^{2^i})\mu\sqrt{n}$$

$$\leq (1 + \delta)\mu\sqrt{n}\sum_{i=0}^{\infty} \delta^{2^i}$$

$$\leq \frac{\delta(1 + \delta)}{1 - \delta}\mu\sqrt{n},$$

where the second inequality follows from (2.22) and the third inequality from (2.21).

□

Note that if $\delta(y, \mu) = 0$ then according to Lemmas 2.4 and 2.5 we have $\phi_\mathrm{B}(y, \mu) = \phi_\mathrm{B}(y(\mu), \mu)$ and $b^T y = b^T y(\mu)$, which is obviously true since $y = y(\mu)$.

2.3.2 Complexity analysis

In this section we will first derive an upper bound for the total number of iterations for the long– and medium–step variants. We will give an upper bound for the number of outer iterations, i.e. the number of updates of the barrier parameter. Then we will give an upper bound for the number of inner iterations during an arbitrary outer iteration. The product of these two bounds is an upper bound for the total number of iterations for the medium– and long–step variant. At the end of this section we will treat the short–step variant. In this case one full Newton step is sufficient to return to the vicinity of the new center.

The following theorem gives an upper bound for the number of outer iterations. In the Logarithmic Barrier Algorithm we will use $\tau = \frac{1}{2}$.

Theorem 2.2 *After at most*

$$\frac{1}{\theta} \ln \frac{4n\mu_0}{\epsilon}$$

outer iterations, the Logarithmic Barrier Algorithm ends up with a dual solution y such that $z^* - b^T y \leq \epsilon$.

Proof: The algorithm stops when $\mu_K = (1 - \theta)^K \mu_0 \leq \frac{\epsilon}{4n}$. Taking logarithms we require

$$-K \ln(1 - \theta) \geq \ln \frac{4n\mu_0}{\epsilon}.$$

Since $\theta \leq -\ln(1 - \theta)$, this certainly holds if

$$K \geq \frac{1}{\theta} \ln \frac{4n\mu_0}{\epsilon}.$$

From (2.11) and Lemma 2.5 we can derive an upper bound for the gap $z^* - b^T y$ after the algorithm has stopped

$$
\begin{aligned}
z^* - b^T y &= z^* - b^T y(\mu_K) + b^T y(\mu_K) - b^T y \\
&\leq \mu_K \left(n + \frac{3}{2}\sqrt{n} \right) \\
&\leq \frac{\epsilon}{4n} \left(n + \frac{3}{2}\sqrt{n} \right) \\
&\leq \epsilon.
\end{aligned}
$$

$\square$

It is well–known that a 2^{-2L}–optimal dual solution can be rounded to an optimal solution of $(\mathcal{LD})$ in $O(n^3)$ arithmetic operations. (See e.g. [120].) Consequently, for this purpose it suffices to take $\epsilon = 2^{-2L}$.

The following lemma is needed to derive an upper bound for the number of inner iterations in each outer iteration. It states that at least a constant decrease in the barrier function value can be obtained by taking a step along the Newton direction.

Lemma 2.6 *Let* $\bar\alpha := (1+\delta)^{-1}$. *Then*

$$
\Delta\phi_B := \phi_B(y,\mu) - \phi_B(y + \bar\alpha p, \mu) \geq \delta - \ln(1+\delta) \geq 0.
$$

Proof: We write down the Taylor series for ϕ_B with respect to α:

$$
\phi_B(y + \alpha p, \mu) = \phi_B(y,\mu) + \alpha g^T p + \frac{1}{2}\alpha^2 p^T H p + \sum_{k=3}^{\infty} t_k,
$$

where t_k denotes the k–th order term in the Taylor series. We will also use the notation a_i for the i–th column of A. Since the k–th order differential, $k \geq 1$, of $-\ln s_i$ at y and p is equal to

$$
(-1)^k \left(\frac{a_i^T p}{s_i} \right)^k (k-1)!,
$$

we have

$$
\begin{aligned}
t_k &= \frac{\alpha^k}{k!} \sum_{i=1}^{n} (-1)^k \left(\frac{a_i^T p}{s_i} \right)^k (k-1)! \\
&= \frac{(-\alpha)^k}{k} \sum_{i=1}^{n} \left(\frac{a_i^T p}{s_i} \right)^k.
\end{aligned}
$$

Hence, we have

$$
\begin{aligned}
|t_k| &\leq \frac{\alpha^k}{k} \sum_{i=1}^{n} \left| \frac{a_i^T p}{s_i} \right|^k \\
&\leq \frac{\alpha^k}{k} \left(\sum_{i=1}^{n} \left| \frac{a_i^T p}{s_i} \right|^2 \right)^{\frac{k}{2}} \\
&= \frac{\alpha^k}{k} \delta^k.
\end{aligned}
$$

So we find, since $g^T p = -p^T H p = -\delta^2$ (Lemma 2.2),

$$\phi_B(y + \alpha p, \mu) \leq \phi_B(y, \mu) + (\frac{1}{2}\alpha^2 - \alpha)\delta^2 + \sum_{k=3}^{\infty} \frac{\alpha^k}{k}\delta^k$$

$$= \phi_B(y, \mu) - \alpha\delta^2 - \ln(1 - \alpha\delta) - \alpha\delta.$$

The last equality only holds if

$$\alpha\delta < 1. \tag{2.22}$$

Hence

$$\Delta\phi_B \geq \alpha(\delta^2 + \delta) + \ln(1 - \alpha\delta). \tag{2.23}$$

The right–hand side is maximal if $\alpha = \bar{\alpha} = (1 + \delta)^{-1}$. This value for α also satisfies condition (2.22). Substitution of this value finally gives

$$\Delta\phi_B \geq \delta - \ln(1 + \delta).$$

This proves the lemma.   □

The following theorem gives an upper bound for the number of inner iterations in each outer iteration.

Theorem 2.3 *Each outer iteration requires at most*

$$\frac{11\theta}{(1 - \theta)^2} \left(\theta n + \frac{3}{2}\sqrt{n}\right) + \frac{11}{3}$$

inner iterations.

Proof: We denote the barrier parameter value in an arbitrary outer iteration by $\bar{\bar{\mu}}$, while the parameter value in the previous outer iteration is denoted by $\bar{\mu}$. The iterate at the beginning of the outer iteration is denoted by y. Hence y is centered with respect to $y(\bar{\mu})$ and $\bar{\bar{\mu}} = (1 - \theta)\bar{\mu}$. Note that because of Lemma 2.6 during each inner iteration the decrease in the barrier function value is at least $\delta - \ln(1 + \delta)$. Since this function is increasing in δ and during each inner iteration we have $\delta \geq \frac{1}{2}$ it follows that the decrease is at least

$$\Delta = \frac{1}{2} - \ln(1 + \frac{1}{2}) > \frac{1}{11}. \tag{2.24}$$

Now let N denote the number of inner iterations during one outer iteration. Since the gap between the barrier function value in the current iterate y and the next center $y(\bar{\bar{\mu}})$ is equal to $\phi_B(y, \bar{\bar{\mu}}) - \phi_B(y(\bar{\bar{\mu}}), \bar{\bar{\mu}})$, we have

$$N\Delta \leq \phi_B(y, \bar{\bar{\mu}}) - \phi_B(y(\bar{\bar{\mu}}), \bar{\bar{\mu}}). \tag{2.25}$$

Let us call the right–hand side of (2.25) $\Phi_B(y, \bar{\bar{\mu}})$, i.e.

$$\Phi_B(y, \bar{\bar{\mu}}) = \phi_B(y, \bar{\bar{\mu}}) - \phi_B(y(\bar{\bar{\mu}}), \bar{\bar{\mu}}).$$

According to the Mean Value Theorem there is a $\hat{\mu} \in (\bar{\bar{\mu}}, \bar{\mu})$ such that

$$\Phi_{\mathrm{B}}(y, \bar{\bar{\mu}}) = \Phi_{\mathrm{B}}(y, \bar{\mu}) + \frac{d\,\Phi_{\mathrm{B}}(y, \mu)}{d\,\mu}\bigg|_{\mu=\hat{\mu}} (\bar{\bar{\mu}} - \bar{\mu}). \tag{2.26}$$

Let us now look at $\frac{d\,\Phi_{\mathrm{B}}(y,\mu)}{d\,\mu}$. We have from (2.8)

$$\frac{d\,\phi_{\mathrm{B}}(y, \mu)}{d\,\mu} = \frac{b^T y}{\mu^2},$$

and, denoting the derivative of $y(\mu)$ with respect to μ by y',

$$\begin{aligned}
\frac{d\,\phi_{\mathrm{B}}(y(\mu), \mu)}{d\,\mu} &= \frac{b^T y(\mu)}{\mu^2} - \frac{b^T y'}{\mu} + \sum_{i=1}^{n} \frac{a_i^T y'}{s_i(\mu)} \\
&= \frac{b^T y(\mu)}{\mu^2} - \frac{b^T y'}{\mu} + e^T S(\mu)^{-1} A^T y' \\
&= \frac{b^T y(\mu)}{\mu^2} - \frac{b^T y'}{\mu} + \frac{(Ax(\mu))^T y'}{\mu} \\
&= \frac{b^T y(\mu)}{\mu^2},
\end{aligned}$$

where the last two equations follow from (2.9). So

$$-\frac{d\,\Phi_{\mathrm{B}}(y, \mu)}{d\,\mu}\bigg|_{\mu=\hat{\mu}} = \frac{b^T y(\mu) - b^T y}{\mu^2}\bigg|_{\mu=\hat{\mu}} \leq \frac{|b^T y(\bar{\bar{\mu}}) - b^T y|}{\bar{\bar{\mu}}^2},$$

where the last inequality follows from the fact that $\bar{\bar{\mu}} < \hat{\mu}$ and from Lemma 2.1. Substituting this into (2.26) gives

$$\begin{aligned}
\Phi_{\mathrm{B}}(y, \bar{\bar{\mu}}) &\leq \Phi_{\mathrm{B}}(y, \bar{\mu}) + \frac{|b^T y(\bar{\bar{\mu}}) - b^T y|}{\bar{\bar{\mu}}^2}(\bar{\mu} - \bar{\bar{\mu}}) \\
&= \Phi_{\mathrm{B}}(y, \bar{\mu}) + \\
&\quad \left(\frac{|b^T y(\bar{\bar{\mu}}) - b^T y|}{\bar{\bar{\mu}}} + \frac{b^T y(\bar{\bar{\mu}}) - b^T y(\bar{\mu})}{\bar{\bar{\mu}}} \right) \frac{(\bar{\mu} - \bar{\bar{\mu}})}{\bar{\bar{\mu}}}. \tag{2.27}
\end{aligned}$$

Because y is centered with respect to $\bar{\mu}$ we have due to Lemma 2.4,

$$\Phi_{\mathrm{B}}(y, \bar{\mu}) \leq \frac{1}{3}.$$

Now note that due to Lemma 2.5 and $\delta(y, \bar{\mu}) \leq \frac{1}{2}$

$$|b^T y(\bar{\mu}) - b^T y| \leq \frac{3}{2}\bar{\mu}\sqrt{n}.$$

Moreover, because of the monotonicity (Lemma 2.1), we have

$$\begin{aligned}
b^T y(\bar{\mu}) - b^T y(\bar{\bar{\mu}}) &\leq b^T y(\bar{\mu}) - c^T x(\bar{\mu}) + c^T x(\bar{\bar{\mu}}) - b^T y(\bar{\bar{\mu}}) \\
&= n(\bar{\mu} - \bar{\bar{\mu}}) \\
&= \theta n \bar{\mu}.
\end{aligned}$$

Plugging all these upper bounds in (2.27) gives

$$
\begin{aligned}
\Phi_{\text{B}}(y,\bar{\mu}) &\leq \frac{1}{3} + \left(\frac{\frac{3}{2}\sqrt{n}}{1-\theta} + \frac{\theta n}{1-\theta} \right) \frac{\theta}{1-\theta} \\
&= \frac{1}{3} + \frac{\theta}{(1-\theta)^2} \left(\frac{3}{2}\sqrt{n} + \theta n \right).
\end{aligned}
\tag{2.28}
$$

Substituting this into (2.25) it follows that

$$
N\Delta \leq \frac{1}{3} + \frac{\theta}{(1-\theta)^2} \left(\frac{3}{2}\sqrt{n} + \theta n \right).
$$

Combining this with (2.24) the theorem follows. $\qquad\square$

Combining Theorems 2.2 and 2.3, the total number of iterations turns out to be given by the following theorem.

Theorem 2.4 *An upper bound for the total number of Newton iterations is given by*

$$
\left[\frac{11}{(1-\theta)^2} \left(\theta n + \frac{3}{2}\sqrt{n} \right) + \frac{11}{30} \right] \ln \frac{4n\mu_0}{\epsilon}.
$$

$\qquad\square$

This makes clear that to obtain an ϵ–optimal solution the algorithm needs

- $O(n \ln \frac{n\mu_0}{\epsilon})$ Newton iterations for the long–step variant $(0 < \theta < 1)$;

- $O(\sqrt{n} \ln \frac{n\mu_0}{\epsilon})$ Newton iterations for the medium–step variant $(\theta = \frac{\nu}{\sqrt{n}}, \nu > 0)$.

To obtain an optimal solution we have to take $\epsilon = 2^{-2L}$. For this value of ϵ the iteration bounds become $O(nL)$ and $O(\sqrt{n}L)$ respectively, assuming that $\mu_0 \leq 2^{O(L)}$.

At the end of each outer iteration we have a dual feasible y such that $\delta(y, \mu) \leq 1$. The following lemma (due to Roos and Vial [133]) shows that a primal feasible solution can be obtained by performing a projection.

Lemma 2.7 *If $\delta := \delta(y, \mu) \leq 1$ then $x(y, \mu)$ is primal feasible. Moreover,*

$$
\mu(n - \delta\sqrt{n}) \leq c^T x(y, \mu) - b^T y \leq \mu(n + \delta\sqrt{n}).
$$

Proof: By the definition of $x(y, \mu)$ we have $Ax(y, \mu) = b$ and

$$
\left\| \frac{Sx(y, \mu)}{\mu} - e \right\| \leq 1.
$$

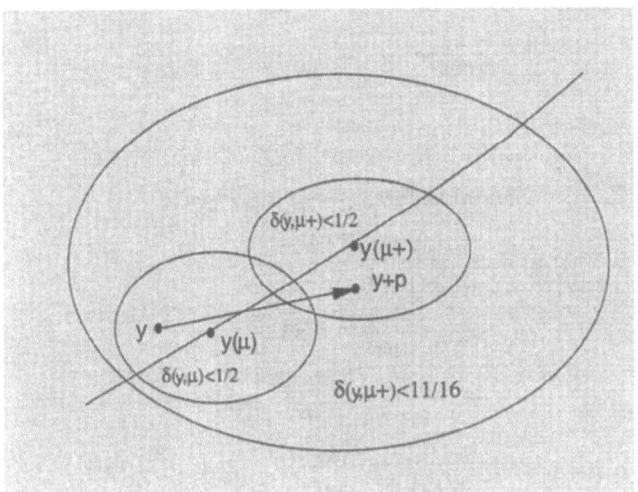

Figure 2.6: One short–step iteration.

This implies $x(y, \mu) \geq 0$, so $x(y, \mu)$ is primal feasible. Moreover, using the Cauchy–Schwartz inequality,

$$
\begin{aligned}
\left| \frac{s^T x(y, \mu)}{\mu} - n \right| &= \left| e^T \left(\frac{S x(y, \mu)}{\mu} - e \right) \right| \\
&\leq \|e\| \, \left\| \frac{S x(y, \mu)}{\mu} - e \right\| \\
&= \delta \sqrt{n}.
\end{aligned}
$$

Since $s^T x(y, \mu) = c^T x(y, \mu) - b^T y$ the lemma follows. □

We want to point out that a complexity analysis can easily be given for the short–step path–following method using some of the lemmas given above. Short–step path–following methods start at a nearly centered iterate and after the parameter is reduced by a small factor, a unit Newton step is taken. The reduction parameter is taken sufficiently small, such that the new iterate is again nearly centered with respect to the new center. Short–step barrier methods for linear programming have been given by Gonzaga [47] and Roos and Vial [133].

Lemma 2.9 states that if θ is small, then we obtain such a short–step path–following method. It shows that after a small reduction of the barrier parameter, one unit Newton step (i.e. the steplength is one) is enough to reach the vicinity of the new reference point on the central path. Theorem 2.2 gives that this short–step algorithm requires $O(\sqrt{n} \ln \frac{n\mu_0}{\epsilon})$ unit Newton steps. The following lemma, proved in Roos and Vial [133], is needed in the proof of Lemma 2.9.

Lemma 2.8 *Let $\mu^+ := (1 - \theta)\mu$. Then we have*

$$\delta(y, \mu^+) \leq \frac{1}{1 - \theta}(\delta(y, \mu) + \theta\sqrt{n}).$$

Proof: Due to the definition of the δ-measure we have

$$
\begin{aligned}
\delta(y, \mu^+) &= \left\| \frac{Sx(y, \mu^+)}{\mu^+} - e \right\| \\
&\leq \left\| \frac{Sx(y, \mu)}{\mu^+} - e \right\| \\
&= \left\| \frac{1}{1 - \theta}\left(\frac{Sx(y, \mu)}{\mu} - e \right) + \left(\frac{1}{1 - \theta} - 1 \right)e \right\| \\
&\leq \frac{1}{1 - \theta}\left\| \frac{Sx(y, \mu)}{\mu} - e \right\| + \frac{\theta}{1 - \theta}\|e\| \\
&= \frac{1}{1 - \theta}(\delta(y, \mu) + \theta\sqrt{n}),
\end{aligned}
$$

where the last equality follows from (2.13). This proves the lemma.   □

Lemma 2.9 *Let $y^+ := y + p$ and $\mu^+ := (1 - \theta)\mu$, where $\theta = \frac{1}{9\sqrt{n}}$. If $\delta(y, \mu) \leq \frac{1}{2}$ then $\delta(y^+, \mu^+) \leq \frac{1}{2}$.*

Proof: Due to Lemma 2.8 we have that

$$\delta(y, \mu^+) \leq \frac{1}{1 - \frac{1}{9\sqrt{n}}}\left(\frac{1}{2} + \frac{1}{9} \right) \leq \frac{11}{16}.$$

Now we can apply the quadratic convergence result (Lemma 2.3)

$$\delta(y^+, \mu^+) \leq \delta(y, \mu^+)^2 \leq \frac{121}{256} < \frac{1}{2}.$$

Figure 2.6 shows the steps in this proof.   □

2.3.3   Illustration of the Newton process

In this subsection we will illustrate the Newton process by means of an example. Consider the following simple linear programming problem:

$$
\begin{cases}
\max y_1 + 2y_2 \\
y_1 \leq 0 \\
y_2 \leq 0.
\end{cases}
$$

It can easily be verified that the central path is given by

$$y(\mu) = -\mu \begin{pmatrix} 1 \\ \frac{1}{2} \end{pmatrix}.$$

The Newton direction (2.14) is given by

$$
\begin{aligned}
p(y,\mu) &= (AS^{-2}A^T)^{-1} \left(\frac{b}{\mu} - AS^{-1}e \right) \\
&= \begin{pmatrix} \frac{y_1^2}{\mu} + y_1 \\ \frac{2y_2^2}{\mu} + y_2 \end{pmatrix},
\end{aligned}
\tag{2.29}
$$

and

$$S^{-1}p_s(y,\mu) := -S^{-1}A^T p(y,\mu) = - \begin{pmatrix} \frac{y_1}{\mu} + 1 \\ \frac{2y_2}{\mu} + 1 \end{pmatrix}.$$

According to (2.18) we have that

$$
\begin{aligned}
\delta(y,\mu)^2 &= \|S^{-1}p_s\|^2 \\
&= (\frac{y_1}{\mu} + 1)^2 + (\frac{2y_2}{\mu} + 1)^2.
\end{aligned}
\tag{2.30}
$$

Consequently, for this example the boundary of the quadratic convergence region $\delta(y,\mu) < 1$ is an ellipsoid and its center is $y(\mu) = (-\mu, -\frac{\mu}{2})$.

Let $\mu = 1$, with $y_1(1) = -1$ and $y_2(1) = -\frac{1}{2}$. Let

$$y^0 = \begin{pmatrix} -\frac{3}{2} \\ -\frac{3}{4} \end{pmatrix}.$$

Since, using (2.30),

$$\delta(y^0, 1) = \sqrt{\frac{1}{4} + \frac{1}{4}} = \sqrt{\frac{1}{2}},$$

we have that y^0 is in the quadratic convergence region of $y(1)$. Let us verify this by taking Newton steps and calculating the δ–measure. Let y^1 and y^2 denote the points obtained by doing one and two full Newton steps respectively. For y^1 we obtain, using (2.29),

$$
\begin{aligned}
y^1 &= y^0 + p(y^0, 1) \\
&= \begin{pmatrix} 2y_1^0 + (y_1^0)^2 \\ 2y_2^0 + 2(y_2^0)^2 \end{pmatrix} \\
&= \begin{pmatrix} -\frac{3}{4} \\ -\frac{3}{8} \end{pmatrix}.
\end{aligned}
$$

Now we can calculate the δ-measure in y^1, using 2.30:

$$\delta(y^1, 1) = \sqrt{\frac{1}{16} + \frac{1}{16}} = \frac{1}{2}\sqrt{\frac{1}{2}}.$$

Note that

$$\delta(y^1, 1) = \delta(y^0, 1)^3 < \delta(y^0, 1)^2,$$

which is in accordance with Lemma 2.3. Let us calculate the next Newton iterate:

$$
\begin{aligned}
y^2 &= y^1 + p(y^1, 1) \\
&= \begin{pmatrix} 2y_1^1 + (y_1^1)^2 \\ 2y_2^1 + 2(y_2^1)^2 \end{pmatrix} \\
&= \begin{pmatrix} -\frac{15}{16} \\ -\frac{15}{32} \end{pmatrix}.
\end{aligned}
$$

Now we calculate the δ-measure in y^2:

$$\delta(y^2, 1) = \sqrt{\frac{1}{256} + \frac{1}{256}} = \frac{1}{8}\sqrt{\frac{1}{2}}.$$

Again, note that

$$\delta(y^2, 1) = \sqrt{\frac{1}{2}}\delta(y^1, 1)^2 < \delta(y^1, 1)^2,$$

which is again in accordance with Lemma 2.3.

Let us now start in a point outside the quadratic convergence region, say $y^0 = (-5, -2)$, for which we have $\delta(y^0, 1) = 5$. Let us take a step α along the Newton direction. So, for the next point y^1 we have

$$
\begin{aligned}
y^1 &= y^0 + \alpha p(y^0, 1) \\
&= \begin{pmatrix} y_1^0 \\ y_2^0 \end{pmatrix} + \alpha \begin{pmatrix} y_1^0 + (y_1^0)^2 \\ y_2^0 + 2(y_2^0)^2 \end{pmatrix} \\
&= \begin{pmatrix} -5 + 20\alpha \\ -2 + 6\alpha \end{pmatrix}.
\end{aligned}
$$

Lemma 2.6 guarantees that the logarithmic barrier function value will decrease by at least $5 - \ln(1+5)$ (which is about 3.21) by taking $\alpha = \frac{1}{6}$. Let us verify this. Since we have

$$\phi_B(y, 1) = -y_1 - 2y_2 - \ln(-y_1) - \ln(-y_2),$$

the difference in logarithmic barrier function value between y^0 and y^1 is

$$\phi_B(y^0, 1) - \phi_B(y^1, 1) = 32\alpha + \ln(1 - 4\alpha) + \ln(1 - 3\alpha). \tag{2.31}$$

Substituting $\alpha = \frac{1}{6}$ we get a decrease of about 3.54, which is indeed larger than the lower bound guaranteed by Lemma 2.6. By differentiating the right–hand side of (2.31) we obtain that the logarithmic barrier function value is minimized for $\alpha = \frac{20}{91}$, giving a reduction of about 3.84. (Observe that to remain strictly feasible α has to satisfy $\alpha < \frac{1}{4}$.)

2.4 Convex quadratic programming

2.4.1 Properties on and near the central path

We consider the convex quadratic programming problem in standard form[7]:

$$(\mathcal{QP}) \qquad \begin{cases} \min q(x) = c^T x + \frac{1}{2} x^T Q x \\ Ax = b, \ x \geq 0. \end{cases}$$

Here Q is a symmetric, positive semi–definite $n \times n$ matrix, A is an $m \times n$ matrix, b and c are $m-$ and $n-$ dimensional vectors respectively; the $n-$dimensional vector x is the variable in which the minimization is carried out. The dual formulation to $(\mathcal{QP})$ is:

$$(\mathcal{QD}) \qquad \begin{cases} \max d(x,y) = b^T y - \frac{1}{2} x^T Q x \\ A^T y - Qx + s = c, \ s \geq 0, \end{cases}$$

where y is an $m-$dimensional vector.

It is well–known that for all x feasible for $(\mathcal{QP})$ and $(\bar{x},y)$ feasible for $(\mathcal{QD})$ we have

$$d(\bar{x},y) \leq z^* \leq q(x),$$

where z^* denotes the optimal objective value for $(\mathcal{QP})$. Optimality holds if and only if the complementary slackness relation $x^T s = 0$ is satisfied, and then $q(x) = d(\bar{x},y)$.

Without loss of generality we assume all coefficients to be integers. We shall denote by L the length of the input data of $(\mathcal{QP})$. We make the standard assumptions that $(\mathcal{QP})$ has a strictly feasible solution, i.e. a feasible solution x such that $x > 0$, and that the feasible set of $(\mathcal{QP})$ is bounded[8]. In order to simplify the analysis we shall also assume that A has full rank, though this assumption is not essential.

We consider the logarithmic barrier function

$$\phi_{\mathrm{B}}(x,\mu) := \frac{c^T x + \frac{1}{2} x^T Q x}{\mu} - \sum_{j=1}^{n} \ln x_j, \qquad (2.32)$$

---

[7]To show how the analysis goes for equality constraints, we will now work with equality constraints.

[8]We think that as in the linear programming case this assumption can be weakened to the boundedness of the optimal set. However, we do not have a rigorous proof for it.

where μ is a positive parameter. The first and second order derivatives of ϕ_B will be denoted by $g = g(x, \mu)$ and $H = H(x, \mu)$, respectively. So

$$g = \nabla \phi_B(x, \mu) = \frac{c + Qx}{\mu} - X^{-1}e,$$

and

$$H = \nabla^2 \phi_B(x, \mu) = \frac{1}{\mu}Q + X^{-2}. \tag{2.33}$$

It can easily be verified that under the assumptions made the logarithmic barrier function has a unique minimum. The necessary and sufficient first order optimality conditions for this unique minimum of $\phi_B(x, \mu)$ are:

$$\begin{cases} A^T y - Qx + s = c, & s \geq 0 \\ Ax = b, & x \geq 0 \\ Xs = \mu e, \end{cases} \tag{2.34}$$

where y and s are $m-$ and $n-$dimensional vectors respectively.

The unique solution of this system is again denoted by $(x(\mu), y(\mu), s(\mu))$. The duality gap in this solution satisfies

$$x(\mu)^T s(\mu) = n\mu, \tag{2.35}$$

according to (2.6). It is well–known that $x(\mu)$ and $y(\mu)$ are continuously differentiable. Hence, if $\mu \downarrow 0$ then $x(\mu)$ and $(x(\mu), y(\mu))$ will converge to optimal primal and dual solutions respectively.

The following lemma states that the primal objective decreases along the primal path and the dual objective increases along the dual path. In the proof we will use the logarithmic barrier function for (QD):

$$\phi_B^d(x, y, \mu) = -\frac{d(x, y)}{\mu} - \sum_{i=1}^n \ln s_i.$$

Lemma 2.10 *The objective $q(x(\mu))$ of the primal problem (QP) is monotonically decreasing and the objective $d(x(\mu), y(\mu))$ of the dual problem (QD) is monotonically increasing if μ decreases.*

Proof: The first part is a classical result of Fiacco and McCormick [35] (see the proof of Lemma 2.1). To prove the second part, first note that $\phi_B^d(x, y, \mu)$ is convex. The Karush–Kuhn–Tucker conditions for a minimizing point are:

$$\begin{cases} A^T y - Qx + s = c, & s \geq 0 \\ \mu AS^{-1}e = b, & x \geq 0 \\ Q(x - \mu S^{-1}e) = 0. \end{cases} \tag{2.36}$$

It is easy to see that the pair $(x(\mu), y(\mu))$ satisfies these conditions. Hence $(x(\mu), y(\mu))$ is a minimizer of $\phi_B^d(x, y, \mu)$. The classical result of Fiacco and McCormick [35] can now be used again to prove the second part of the lemma. (See also Anstreicher [4].) □

To measure the distance to the central path of non–centered points, we will introduce three proximity measures. The first is analogous to Roos and Vial's [133] appealing measure (2.12) for linear programming

$$\delta(x, \mu) := \min_y \|\frac{X}{\mu}(c + Qx - A^T y) - e\|. \tag{2.37}$$

In the sequel of this section we will sometimes write δ instead of $\delta(x, \mu)$ for briefness' sake. Loosely speaking, $\delta(x, \mu)$ measures the deviation from optimality condition (2.34). The unique solution of the minimization problem in the definition of $\delta(x, \mu)$ is denoted by $y(x, \mu)$ and the corresponding slack variable by $s(x, \mu)$ (i.e. $s(x, \mu) = c + Qx - A^T y(x, \mu)$). Hence $\delta(x, \mu)$ can be written as

$$\delta(x, \mu) = \|\frac{X}{\mu}(c + Qx - A^T y(x, \mu)) - e\|. \tag{2.38}$$

It can easily be verified that

$$x = x(\mu) \iff \delta(x, \mu) = 0,$$

and moreover

$$\delta(x, \mu) = 0 \implies y(x, \mu) = y(\mu).$$

In the algorithm approximate line searches along projected Newton directions are carried out to minimize ϕ_B for fixed μ, i.e. the directions correspond to exact minimization of the quadratic approximation to ϕ_B on the affine space $Ax = b$. This means that the Newton direction $p = p(x, \mu)$ is determined by solving

$$g + Hp = A^T \bar{y}, \quad Ap = 0, \tag{2.39}$$

as a linear system of equations[9] in p and $\bar{y}$. Gill et al. [41] give two alternative (equivalent) forms for this direction:

- Range–space form.

  From the first equation in (2.39) we get

$$p = H^{-1} A^T \bar{y} - H^{-1} g. \tag{2.40}$$

  Substituting this into $Ap = 0$ we obtain $\bar{y} = (AH^{-1}A^T)^{-1}AH^{-1}g$. This can be substituted into (2.40) to obtain the range–space form of $p(x, \mu)$

$$\begin{aligned} p(x, \mu) &= -H^{-1}(I - A^T(AH^{-1}A^T)^{-1}AH^{-1})g \\ &= -H^{-\frac{1}{2}} P_{AH^{-\frac{1}{2}}} H^{-\frac{1}{2}} g, \end{aligned} \tag{2.41}$$

  where $P_{AH^{-\frac{1}{2}}}$ denotes the orthogonal projection onto the null–space of $AH^{-\frac{1}{2}}$;

---

[9]In fact the problem which has to be solved to obtain p is: $\min \phi_B(x, \mu) + g^T p + \frac{1}{2} p^T Hp$, with $Ap = 0$, where x is the current iterate.

- Null–space form.

Let Z be an $n \times (n-m)$ matrix, with independent columns, such that $AZ = 0$, i.e. the columns of Z are a basis for the null–space of A. Now, equations (2.39) are equivalent with

$$p = Zr, \ g + HZr = A^T \bar{y}.$$

Multiplying the last equation with Z^T we obtain

$$p = Zr, \ Z^T g + Z^T HZr = 0.$$

From these equations it is easy to obtain the null–space form of $p(x, \mu)$

$$p(x, \mu) = -Z(Z^T HZ)^{-1} Z^T g. \tag{2.42}$$

The second and third measure for the distance to the central path are $\|X^{-1} p(x, \mu)\|$ and $\|p(x, \mu)\|_{H(x,\mu)}$, where the latter is defined by

$$\|p(x, \mu)\|^2_{H(x,\mu)} = p(x, \mu)^T H(x, \mu) p(x, \mu).$$

Although all of the three measures will be used in the analysis, only $\|p\|_H$ is used in the algorithm. We note that $\|.\|_{H(x,\mu)}$ defines a norm, because $H(x, \mu)$ is positive definite.

We will work with the null–space form for p, because it facilitates the analysis a lot. In the analysis we will also assume that $(X^{-1} Z)^T (X^{-1} Z) = I$, hence $X^{-1} Z$ is orthonormal. In this case we have the following well–known properties:

Property 1. $(X^{-1} Z)(X^{-1} Z)^T$ is the projection onto the null–space of AX, since the null–space of AX is the range–space of $X^{-1} Z$.

Property 2. $\|X^{-1} Z\lambda\| = \|\lambda\|$ for any λ, since

$$\|X^{-1} Z\lambda\|^2 = \lambda^T (X^{-1} Z)^T X^{-1} Z\lambda = \|\lambda\|^2.$$

The last equality follows because $X^{-1} Z$ is orthonormal.

Property 3. $\|(X^{-1} Z)^T \nu\| \leq \|\nu\|$ for any ν, with equality if ν is in the null–space of AX. This can be explained as follows:

$$\|(X^{-1} Z)^T \nu\|^2 = \nu^T (X^{-1} Z)(X^{-1} Z)^T \nu \leq \|\nu\|,$$

where the last inequality follows because $(X^{-1} Z)(X^{-1} Z)^T$ is a projection matrix (Property 1). This also explains that equality holds if ν is in the null–space of AX.

The following lemma shows that there is a close connection between the three measures.

Lemma 2.11 *For given x and μ we have*

$$\|X^{-1} p\|^2 \leq \|p\|^2_H = -p^T g \leq \delta^2$$

Proof: We may write, using (2.33),

$$
\begin{aligned}
\|p\|_H^2 &= p^T H p \\
&= p^T (X^{-2} + \frac{1}{\mu} Q) p \\
&\geq \|X^{-1} p\|^2.
\end{aligned}
$$

This proves the first inequality. The equality of the lemma follows simply from (2.39)

$$
\begin{aligned}
\|p\|_H^2 &= p^T H p \\
&= p^T (A^T \bar{y} - g) \\
&= -p^T g.
\end{aligned}
$$

So, it remains to prove the last inequality of the lemma. Using the definition of p (2.42), and the fact that $X^{-1} Z$ is orthonormal, it follows that

$$
\begin{aligned}
-g^T p &= g^T Z (Z^T H Z)^{-1} Z^T g \\
&= g^T Z \left(Z^T X^{-2} Z + \frac{1}{\mu} Z^T Q Z \right)^{-1} Z^T g \\
&= g^T Z \left(I + \frac{1}{\mu} Z^T Q Z \right)^{-1} Z^T g \\
&\leq \|Z^T g\|^2 \| \left(I + \frac{1}{\mu} Z^T Q Z \right)^{-1} \| \\
&\leq \|Z^T g\|^2,
\end{aligned}
\tag{2.43}
$$

where the last inequality follows because the eigenvalues of $\left(I + \frac{1}{\mu} Z^T Q Z \right)^{-1}$ are all less than or equal to 1. Moreover,

$$
\begin{aligned}
Z^T g &= Z^T \left(\frac{c + Q x}{\mu} - X^{-1} e \right) \\
&= Z^T \left(\frac{c + Q x - A^T y}{\mu} - X^{-1} e \right) \\
&= (X^{-1} Z)^T \left(\frac{X}{\mu} (c + Q x - A^T y) - e \right),
\end{aligned}
\tag{2.44}
$$

where the second equality holds for any y, because $AZ = 0$. From (2.44) and Property 3 it now follows that

$$
\begin{aligned}
\|Z^T g\| &= \|(X^{-1} Z)^T \left(\frac{X}{\mu} (c + Q x - A^T y) - e \right) \| \\
&\leq \| \frac{X}{\mu} (c + Q x - A^T y) - e \|.
\end{aligned}
$$

Putting y equal to $y(x, \mu)$, we obtain

$$\|Z^T g\| \leq \|\frac{X}{\mu}(c + Qx - A^T y(x, \mu)) - e\| = \delta, \qquad (2.45)$$

where the last equality follows from (2.38). Finally, from (2.43) and (2.45) it follows that $-g^T p \leq \delta^2$. This proves the lemma. $\quad\square$

Note that in the LP case, for which $Q = 0$, the projected Newton direction (2.42) reduces to $p = -ZX^{-2}Z^T g$, which coincides with the scaled projected gradient direction. For this case, it is easy to verify that the three measures $\|X^{-1}p\|$, $\|p\|_H$ and δ are equivalent (in Lemma 2.11 equalities hold instead of inequalities; this is in fact Lemma 2.2). Consequently, if we substitute $Q = 0$ in the analysis given below, we will get back the results for LP, given in Section 2.3.

Now we will prove some fundamental lemmas for nearly centered points. The following lemma shows quadratic convergence in the vicinity of the central path for all of the three measures.

Lemma 2.12 *If $\|X^{-1}p(x, \mu)\| < 1$ then $x^+ = x + p(x, \mu)$ is a strictly feasible point for $(\mathcal{QP})$. Moreover,*

$$\|(X^+)^{-1}p(x^+, \mu)\| \leq \|p(x^+, \mu)\|_{H(x^+,\mu)} \leq \delta(x^+, \mu) \leq$$
$$\leq \|X^{-1}p(x, \mu)\|^2 \leq \|p(x, \mu)\|^2_{H(x,\mu)} \leq \delta(x, \mu)^2.$$

Proof:[10] It is easy to see that $Ax^+ = b$, since $Ap(x, \mu) = 0$. Moreover,

$$x^+ = x + p = X(e + X^{-1}p) > 0,$$

because $\|X^{-1}p\| < 1$. This proves the first part of the lemma.

Using the definition of $\delta(x^+, \mu)$ we have

$$\begin{aligned}
\delta(x^+, \mu) &= \min_y \|\frac{X^+}{\mu}\left(c + Qx^+ - A^T y\right) - e\| \\
&= \min_y \|\frac{(X + P)}{\mu}\left(c + Q(x + p) - A^T y\right) - e\|. \qquad (2.46)
\end{aligned}$$

Now using $y = \mu \bar{y}$ in (2.46), where $\bar{y}$ is defined by (2.39), it follows that

$$\begin{aligned}
\delta(x^+, \mu) &\leq \|\frac{(X + P)}{\mu}\left(c + Q(x + p) - \mu A^T \bar{y}\right) - e\| \\
&= \|(X + P)(X^{-1}e - X^{-2}p) - e\| \\
&= \|PX^{-2}p\| \\
&\leq \|X^{-1}p\|^2.
\end{aligned}$$

---

[10]Thanks are due to M.J. Todd who simplified an earlier version of the proof.

This proves the middle inequality of the lemma. The rest follows immediately from Lemma 2.11.   □

The following lemma gives an upper bound for the difference in barrier function value in a nearly centered point x and the center $x(\mu)$.

Lemma 2.13 *If* $\|p\|_H < 1$ *then*

$$\phi_B(x,\mu) - \phi_B(x(\mu),\mu) \le \frac{\|p\|_H^2}{1 - \|p\|_H^2}.$$

Proof: The barrier function ϕ_B is convex for fixed μ, whence

$$\phi_B(x,\mu) - \phi_B(x+p,\mu) \le -g^T p = \|p\|_H^2, \qquad (2.47)$$

where the equality follows from Lemma 2.11. Now, let $x^0 := x$ and let $x^1, x^2, \cdots$ denote the sequence of points obtained by repeating Newton steps, starting at x^0. By Lemma 2.12 we have

$$\|p(x^i,\mu)\|_{H(x^i,\mu)} \le \|p(x^0,\mu)\|_{H(x^0,\mu)}^{2^i}\| = \|p\|_H^{2^i}. \qquad (2.48)$$

This means that $\|p(x^i,\mu)\|_{H(x^i,\mu)} \to 0$ if $i \to \infty$. Since all x^i lie in a bounded region and $\|p(x,\mu)\|_{H(x,\mu)} = 0$ if and only if $x = x(\mu)$, we have that all limit points of the sequence are $x(\mu)$. Hence, the sequence of Newton iterates converges to $x(\mu)$. Consequently, we may write,

$$
\begin{aligned}
\phi_B(x,\mu) - \phi_B(x(\mu),\mu) &= \sum_{i=0}^{\infty} \left(\phi_B(x^i,\mu) - \phi_B(x^{i+1},\mu) \right) \\
&\le \sum_{i=0}^{\infty} \|p(x^i,\mu)\|_{H(x^i,\mu)}^2 \\
&\le \sum_{i=0}^{\infty} \|p\|_H^{2^{i+1}} \\
&\le \frac{\|p\|_H^2}{1 - \|p\|_H^2},
\end{aligned}
$$

where the first inequality follows from (2.47) and the second inequality from (2.48). This proves the lemma.   □

The following lemma gives an upper bound for the difference in objective value in a nearly centered point x and the center $x(\mu)$.

Lemma 2.14 *If* $\|p\|_H < 1$, *then*

$$|q(x) - q(x(\mu))| \le \frac{\|p\|_H(1 + \|p\|_H)}{1 - \|p\|_H}\mu\sqrt{n}.$$

Proof: Since $q(x)$ is convex, we have

$$\nabla q(x)^T p \leq q(x+p) - q(x) \leq \nabla q(x+p)^T p. \tag{2.49}$$

Using that $\nabla q(x) = c + Qx = \mu g + \mu X^{-1}e$, we can derive the following lower bound for the left–hand side of (2.49)

$$\begin{aligned}
\nabla q(x)^T p &= \mu g^T p + \mu e^T X^{-1} p \\
&\geq -\mu \|p\|_H^2 - \mu \|p\|_H \sqrt{n} \\
&\geq -\|p\|_H (1 + \|p\|_H) \mu \sqrt{n},
\end{aligned} \tag{2.50}$$

where the first inequality follows from Lemma 2.11 and the Cauchy–Schwartz inequality. Now, using (2.39), we derive an upper bound for the right–hand side expression in (2.49):

$$\begin{aligned}
\nabla q(x+p)^T p &= c^T p + p^T Q(x+p) \\
&= c^T p + p^T(\mu A^T \bar{y} + \mu X^{-1} e - c - \mu X^{-2} p) \\
&= \mu e^T X^{-1} p - \mu \|X^{-1} p\|^2 \\
&\leq \mu e^T X^{-1} p \\
&\leq \mu \sqrt{n} \|p\|_H.
\end{aligned} \tag{2.51}$$

Consequently, substitution of (2.50) and (2.51) into (2.49) yields

$$|q(x) - q(x+p)| \leq \|p\|_H (1 + \|p\|_H) \mu \sqrt{n}. \tag{2.52}$$

The remainder of the proof follows by considering a sequence of Newton steps initiated at $x^0 := x$, as in the proof of Lemma 2.13. (See also Lemma 2.5.) Let $x^1, x^2, \cdots$ denote the Newton sequence. As shown in the proof of Lemma 2.13 this sequence converges to $x(\mu)$, and moreover (2.48)

$$\|p(x^i, \mu)\|_{H(x^i, \mu)} \leq \|p(x^0, \mu)\|_{H(x^0, \mu)}^{2^i}\| = \|p\|_H^{2^i}. \tag{2.53}$$

Hence, we may write

$$\begin{aligned}
|q(x) - q(x(\mu))| &= \left| \sum_{i=0}^{\infty} q(x^i) - q(x^{i+1}) \right| \\
&\leq \sum_{i=0}^{\infty} \left| q(x^i) - q(x^{i+1}) \right| \\
&\leq \sum_{i=0}^{\infty} \|p(x^i, \mu)\|_{H(x^i,\mu)} (1 + \|p(x^i, \mu)\|_{H(x^i,\mu)}) \mu \sqrt{n} \\
&\leq (1 + \|p\|_H) \mu \sqrt{n} \sum_{i=0}^{\infty} \|p\|_H^{2^i} \\
&\leq \frac{\|p\|_H (1 + \|p\|_H)}{1 - \|p\|_H} \mu \sqrt{n},
\end{aligned}$$

where the second inequality follows from (2.52) and the third inequality from (2.53). This proves the lemma. □

Note that the last two lemmas are analogous to Lemmas 2.4 and 2.5 for the linear case, respectively.

2.4.2 Complexity analysis

In this section we will derive upper bounds for the total number of outer and inner iterations needed by the Logarithmic Barrier Algorithm. First we derive upper bounds for the long– and medium–step variants, and finally for the short–step variant. In the Logarithmic Barrier Algorithm we will take $\tau = \frac{1}{2}$. (Of course, some obvious changes have to be made in this algorithm given in Section 2.1, e.g. instead of y we now have x.)

Theorem 2.5 *After at most*

$$\frac{1}{\theta} \ln \frac{4n\mu_0}{\epsilon}$$

outer iterations, the Logarithmic Barrier Algorithm ends up with a primal solution such that $q(x) - z^* \leq \epsilon$.

Proof: The proof is similar as the proof given for the linear programming case (Theorem 2.2). However, for completeness we will give the proof.

The algorithm stops when $\mu_K = (1 - \theta)^K \mu_0 \leq \frac{\epsilon}{4n}$. Taking logarithms we require

$$-K \ln(1 - \theta) \geq \ln \frac{4n\mu_0}{\epsilon}.$$

Since $\theta \leq -\ln(1 - \theta)$, this certainly holds if

$$K \geq \frac{1}{\theta} \ln \frac{4n\mu_0}{\epsilon}.$$

From Lemma 2.14 and (2.35) we can derive an upper bound for the gap $q(x) - z^*$ after the algorithm has stopped

$$
\begin{aligned}
q(x) - z^* &= q(x) - q(x(\mu_K)) + q(x(\mu_K)) - z^* \\
&\leq \mu_K(\frac{3}{2}\sqrt{n} + n) \\
&\leq \frac{\epsilon}{4n}(\frac{3}{2}\sqrt{n} + n) \\
&\leq \epsilon.
\end{aligned}
$$

$\square$

We note, that this final primal solution can be rounded to an optimal solution for $(\mathcal{QP})$ in $O(n^3)$ arithmetic operations. (See e.g. [120].)

The following lemma is needed to derive an upper bound for the number of inner iterations in each outer iteration. It states that a sufficient decrease in the value of the barrier function can be obtained by taking a step along the Newton direction.

Lemma 2.15 *Let $\bar{\alpha} := (1 + \|p\|_H)^{-1}$. Then*

$$\Delta\phi_{\mathrm{B}} := \phi_{\mathrm{B}}(x, \mu) - \phi_{\mathrm{B}}(x + \bar{\alpha}p, \mu) \geq \|p\|_H - \ln(1 + \|p\|_H).$$

Proof: We write down the Taylor series for ϕ_B with respect to α:

$$\phi_\text{B}(x + \alpha p, \mu) = \phi_\text{B}(x, \mu) + \alpha g^T p + \frac{1}{2}\alpha^2 p^T H p + \sum_{k=3}^{\infty} t_k,$$

where t_k denotes the k–order term in the Taylor series. Since the k–th order differential of $-\ln x_i$ at x and p is equal to

$$(-1)^k \left(\frac{p_i}{x_i}\right)^k (k-1)!,$$

we have

$$
\begin{aligned}
t_k &= \frac{\alpha^k}{k!} \sum_{i=1}^{n} (-1)^k \left(\frac{p_i}{x_i}\right)^k (k-1)! \\
&= \frac{(-\alpha)^k}{k} \sum_{i=1}^{n} \left(\frac{p_i}{x_i}\right)^k.
\end{aligned}
$$

Hence we find, by using Lemma 2.11,

$$
\begin{aligned}
|t_k| &\leq \frac{\alpha^k}{k} \sum_{i=1}^{n} \left(\frac{|p_i|}{x_i}\right)^k \\
&\leq \frac{\alpha^k}{k} \left(\sum_{i=1}^{n} \left(\frac{p_i}{x_i}\right)^2\right)^{\frac{k}{2}} \\
&= \frac{\alpha^k}{k} \|X^{-1}p\|^k \\
&\leq \frac{\alpha^k}{k} \|p\|_H^k.
\end{aligned}
$$

Using Lemma 2.11, we have for the linear and quadratic term in the Taylor series

$$\alpha p^T g + \frac{1}{2}\alpha^2 p^T H p = (\frac{1}{2}\alpha^2 - \alpha)\|p\|_H^2.$$

So we find

$$
\begin{aligned}
\phi_\text{B}(x + \alpha p, \mu) &\leq \phi_\text{B}(x, \mu) + (\frac{1}{2}\alpha^2 - \alpha)\|p\|_H^2 + \sum_{k=3}^{\infty} \frac{\alpha^k}{k} \|p\|_H^k \\
&- \phi_\text{B}(x, \mu) \quad \alpha\|p\|_H^2 - \ln(1 - \alpha\|p\|_H) - \alpha\|p\|_H.
\end{aligned}
$$

Hence

$$\Delta\phi_\text{B} \geq \alpha(\|p\|_H^2 + \|p\|_H) + \ln(1 - \alpha\|p\|_H). \qquad (2.54)$$

The right–hand side is maximal if $\alpha = \bar{\alpha} = (1 + \|p\|_H)^{-1}$. Substitution of this value finally gives

$$\Delta\phi_\text{B} \geq \|p\|_H - \ln(1 + \|p\|_H).$$

This proves the lemma. $\qquad\qquad\qquad\qquad\qquad\qquad\qquad\qquad\qquad\qquad\square$

The following theorem gives an upper bound for the total number of inner iterations in each outer iteration.

Theorem 2.6 *Each outer iteration requires at most*

$$\frac{11\theta}{(1-\theta)^2}\left(\theta n + \frac{3}{2}\sqrt{n}\right) + \frac{11}{3}$$

inner iterations.

Proof: We denote the barrier parameter value in an arbitrary outer iteration by $\bar{\bar{\mu}}$, while the parameter value in the previous outer iteration is denoted by $\bar{\mu}$. The iterate at the beginning of the outer iteration is denoted by x. Hence x is centered with respect to $x(\bar{\mu})$ and $\bar{\bar{\mu}} = (1-\theta)\bar{\mu}$. Note that because of Lemma 2.15 the decrease in the barrier function value is at least $\|p\|_H - \ln(1+\|p\|_H)$ in each inner iteration. Since this function is increasing in $\|p\|_H$ and during each inner iteration we have $\|p\|_H \geq \frac{1}{2}$ it follows that the decrease is at least

$$\Delta = \frac{1}{2} - \ln(1+\frac{1}{2}) > \frac{1}{11}. \tag{2.55}$$

Now let N denote the number of inner iterations during one outer iteration. Since the gap between the barrier function value in the current iterate x and the next center $x(\bar{\bar{\mu}})$ is equal to

$$\phi_B(x,\bar{\bar{\mu}}) - \phi_B(x(\bar{\bar{\mu}}),\bar{\bar{\mu}})$$

we have

$$N\Delta \leq \phi_B(x,\bar{\bar{\mu}}) - \phi_B(x(\bar{\bar{\mu}}),\bar{\bar{\mu}}). \tag{2.56}$$

Let us call the right–hand side of (2.56) $\Phi_B(x,\bar{\bar{\mu}})$, i.e.

$$\Phi_B(x,\bar{\bar{\mu}}) = \phi_B(x,\bar{\bar{\mu}}) - \phi_B(x(\bar{\bar{\mu}}),\bar{\bar{\mu}}).$$

According to the Mean Value Theorem there exists a $\hat{\mu} \in (\bar{\bar{\mu}}, \bar{\mu})$ such that

$$\Phi_B(x,\bar{\bar{\mu}}) = \Phi_B(x,\bar{\mu}) + \left.\frac{d\,\Phi_B(x,\mu)}{d\,\mu}\right|_{\mu=\hat{\mu}} (\bar{\bar{\mu}} - \bar{\mu}). \tag{2.57}$$

Let us now look at $\frac{d\,\Phi_B(x,\mu)}{d\,\mu}$. We have from (2.32)

$$\frac{d\,\phi_B(x,\mu)}{d\,\mu} = -\frac{q(x)}{\mu^2},$$

and, denoting the derivative of $x(\mu)$ with respect to μ by x',

$$\begin{aligned}
\frac{d\,\phi_B(x(\mu),\mu)}{d\,\mu} &= -\frac{q(x(\mu))}{\mu^2} + \frac{\nabla q(x(\mu))^T x'}{\mu} - e^T X^{-1} x' \\
&= -\frac{q(x(\mu))}{\mu^2} + \frac{(c + Qx(\mu))^T x'}{\mu} - \frac{s(\mu)^T x'}{\mu} \\
&= -\frac{q(x(\mu))}{\mu^2},
\end{aligned}$$

where the last two equations follow from (2.34) and from $Ax' = 0$. So

$$
-\frac{d\,\Phi_{\mathrm{B}}(x,\mu)}{d\,\mu}\bigg|_{\mu=\hat{\mu}} = \frac{q(x)-q(x(\mu))}{\mu^2}\bigg|_{\mu=\hat{\mu}}
$$

$$
\leq \frac{|q(x)-q(x(\bar{\mu}))|}{\bar{\mu}^2},
$$

where the last inequality follows from the fact that $\bar{\mu} < \hat{\mu}$ and from Lemma 2.10. Substituting this into (2.57) gives

$$
\Phi_{\mathrm{B}}(x,\bar{\bar{\mu}}) \leq \Phi_{\mathrm{B}}(x,\bar{\mu}) + \frac{|q(x)-q(x(\bar{\mu}))|}{\bar{\mu}^2}(\bar{\mu}-\bar{\bar{\mu}})
$$

$$
= \Phi_{\mathrm{B}}(x,\bar{\mu}) +
$$

$$
\left(\frac{|q(x)-q(x(\bar{\mu}))|}{\bar{\mu}} + \frac{q(x(\bar{\mu}))-q(x(\bar{\bar{\mu}}))}{\bar{\mu}}\right)\frac{(\bar{\mu}-\bar{\bar{\mu}})}{\bar{\mu}}. \qquad (2.58)
$$

Because $\|p(x,\bar{\mu})\|_{H(x,\bar{\mu})} \leq \frac{1}{2}$, we have due to Lemma 2.13

$$
\Phi_{\mathrm{B}}(x,\bar{\mu}) \leq \frac{1}{3}.
$$

Now note that due to Lemma 2.14

$$
|q(x)-q(x(\bar{\mu}))| \leq \frac{3}{2}\bar{\mu}\sqrt{n}.
$$

Moreover, because of the monotonicity property (Lemma 2.10), we have

$$
q(x(\bar{\mu}))-q(x(\bar{\bar{\mu}})) \leq q(x(\bar{\mu})) - d(x(\bar{\mu}),y(\bar{\mu}))
$$

$$
+d(x(\bar{\bar{\mu}}),y(\bar{\bar{\mu}})) - q(x(\bar{\bar{\mu}}))
$$

$$
= n(\bar{\mu}-\bar{\bar{\mu}})
$$

$$
= \theta n\bar{\mu}.
$$

Plugging all these upper bounds in (2.58) gives

$$
\Phi_{\mathrm{B}}(x,\bar{\bar{\mu}}) \leq \frac{1}{3} + \left(\frac{\frac{3}{2}\sqrt{n}}{1-\theta} + \frac{\theta n}{1-\theta}\right)\frac{\theta}{1-\theta}
$$

$$
= \frac{1}{3} + \frac{\theta}{(1-\theta)^2}\left(\frac{3}{2}\sqrt{n}+\theta n\right)
$$

Substituting this into (2.56) it follows that

$$
N\Delta \leq \frac{1}{3} + \frac{\theta}{(1-\theta)^2}\left(\frac{3}{2}\sqrt{n}+\theta n\right).
$$

Combining this with (2.55) the theorem follows.

$\square$

Combining Theorems 2.5 and 2.6, the total number of iterations turns out to be given by the following theorem.

Theorem 2.7 *An upper bound for the total number of Newton iterations is given by*

$$\left[\frac{11}{(1-\theta)^2}\left(\theta n + \frac{3}{2}\sqrt{n}\right) + \frac{11}{3\theta}\right] \ln \frac{4n\mu_0}{\epsilon}.$$

$\square$

Note that this upper bound is exactly the same as the upper bound for the linear case (Theorem 2.4). Theorem 2.7 makes clear that to obtain an ϵ–optimal solution the algorithm needs

- $O(n \ln \frac{n\mu_0}{\epsilon})$ Newton iterations for the long–step variant $(0 < \theta < 1)$;

- $O(\sqrt{n} \ln \frac{n\mu_0}{\epsilon})$ Newton iterations for the medium–step variant $(\theta = \frac{\nu}{\sqrt{n}}, \nu > 0)$.

To obtain an optimal solution we have to take $\epsilon = 2^{-2L}$. For this value of ϵ the iteration bounds become $O(nL)$ and $O(\sqrt{n}L)$ respectively, assuming that $\mu_0 \leq 2^{O(L)}$.

At the end of each sequence of inner iterations we have a primal feasible x such that $\|p\|_H \leq 1$. The following lemma shows that a dual feasible solution can be obtained by performing an additional full Newton step, and projection.

Lemma 2.16 *Let $x^+ = x + p(x,\mu)$. If $\|p(x,\mu)\|_{H(x,\mu)} \leq 1$ then*

$$\delta := \delta(x^+,\mu) \leq 1$$

and $y := y(x^+,\mu)$ is dual feasible. Moreover, the duality gap satisfies

$$\mu(n - \delta\sqrt{n}) \leq q(x^+) - d(x^+,y) \leq \mu(n + \delta\sqrt{n}).$$

Proof: By Lemma 2.12 we have

$$\delta(x^+,\mu) \leq \|p(x,\mu)\|^2_{H(x,\mu)} \leq 1.$$

By the definition of $s(x,\mu) = c + Qx - A^T y(x,\mu)$ we have

$$\delta(x^+,\mu) = \left\|\frac{X^+ s(x^+,\mu)}{\mu} - e\right\| \leq 1.$$

This implies $s(x^+,\mu) \geq 0$, so $y(x^+,\mu)$ is dual feasible. Moreover,

$$\left|\frac{(x^+)^T s(x^+,\mu)}{\mu} - n\right| = \left|e^T\left(\frac{X^+ s(x^+,\mu)}{\mu} - e\right)\right|$$

$$\leq \|e\| \left\|\frac{X^+ s(x^+,\mu)}{\mu} - e\right\|$$

$$= \delta\sqrt{n}.$$

Consequently, using that $(x^+)^T s(x^+, \mu) = q(x^+) - d(x^+, y)$, we get

$$\mu(n - \delta\sqrt{n}) \leq q(x^+) - d(x^+, y) \leq \mu(n + \delta\sqrt{n}).$$

$\square$

Short–step path–following methods start at a nearly centered iterate and take a unit Newton step after the parameter has been reduced by a small factor. The reduction parameter is sufficiently small, such that the new iterate is again nearly centered with respect to the new center. Theorem 2.5 implies that this short–step algorithm requires $O(\sqrt{n} \ln \frac{n\mu_0}{\epsilon})$ unit Newton steps. Short–step barrier methods for convex quadratic programming have been given by Ye [158], Goldfarb and Liu [44], and Ben Daya and Shetty [12]. Lemma 2.18 shows that if θ is small, we obtain such a short–step path–following method. The proof is analogous to the proof for the linear case (Lemma 2.9). The following lemma will be needed in the proof of Lemma 2.18.

Lemma 2.17 Let $\mu^+ := (1 - \theta)\mu$. Then

$$\delta(x, \mu^+) \leq \frac{1}{1 - \theta}(\delta(x, \mu) + \theta\sqrt{n}).$$

Proof: Due to the definition of our δ–measure we have

$$
\begin{aligned}
\delta(x, \mu^+) &= \left\| \frac{Xs(x, \mu^+)}{\mu^+} - e \right\| \\
&\leq \left\| \frac{Xs(x, \mu)}{\mu^+} - e \right\| \\
&= \left\| \frac{1}{1 - \theta}(\frac{Xs(x, \mu)}{\mu} - e) + (\frac{1}{1 - \theta} - 1)e \right\| \\
&\leq \frac{1}{1 - \theta} \left\| \frac{Xs(x, \mu)}{\mu} - e \right\| + \frac{\theta}{1 - \theta} \|e\| \\
&\leq \frac{1}{1 - \theta}(\delta(x, \mu) + \theta\sqrt{n}),
\end{aligned}
$$

where the last inequality follows from (2.38). This proves the lemma. $\square$

Lemma 2.18 Let $x^+ := x + p(x, \mu)$ and $\mu^+ := (1 - \theta)\mu$, where $\theta = \frac{1}{9\sqrt{n}}$. If $\delta(x, \mu) \leq \frac{1}{2}$ then $\delta(x^+, \mu^+) \leq \frac{1}{2}$.

Proof: Due to Lemma 2.17 we have

$$\delta(x, \mu^+) \leq \frac{1}{1 - \frac{1}{9\sqrt{n}}}(\frac{1}{2} + \frac{1}{9}) \leq \frac{11}{16}.$$

Now we can apply the quadratic convergence result (Lemma 2.12)

$$\delta(x^+, \mu^+) \leq \delta(x, \mu^+)^2 \leq \frac{121}{256} < \frac{1}{2}.$$

$\square$

2.5 Smooth convex programming

2.5.1 On the monotonicity of the primal and dual objectives along the central path

Let us first recall the primal convex programming problem

$$(\mathcal{CP}) \quad \begin{cases} \max\ f_0(y) \\ f_i(y) \le 0,\ i = 1, \cdots, n \\ y \in \mathbb{R}^m, \end{cases}$$

and its dual problem

$$(\mathcal{CD}) \quad \begin{cases} \min\ f_0(y) - \sum_{i=1}^n x_i f_i(y) \\ \sum_{i=1}^n x_i \nabla f_i(y) = \nabla f_0(y) \\ x_i \ge 0, \end{cases}$$

as given in Section 2.1.

Fiacco and McCormick [35] proved that the primal objective is monotonically increasing along the barrier path. For linear and quadratic programming it turned out to be easy to prove that the dual objective is decreasing along the logarithmic barrier path: the dual feasible point $x(\mu)$ associated with $y(\mu)$ is the unique minimum for the dual barrier function, which is also convex. So, again applying Fiacco and McCormick's argument for the dual barrier function directly gives the result. (This idea was used in the proofs of the Lemmas 2.1 and 2.10.) We note that Fiacco and McCormick's monograph does not deal with the monotonicity of the dual objective along barrier paths[11].

In Den Hertog et al. [29] we proved that the dual objective of a convex program is decreasing along the central path defined by the logarithmic barrier function. This was carried out by differentiating the Karush–Kuhn–Tucker conditions and manipulating these equations. In that proof we needed the objective and constraint functions to be twice continuously differentiable.

In this subsection we will prove the same result under weaker assumptions. The assumptions which we will make in the sequel of this subsection are that the functions $-f_0(y)$ and $f_i(y)$, $i = 1, \cdots, n$, are convex and continuously differentiable, and that the feasible region of $(\mathcal{CP})$ is bounded and has a nonempty interior.

We introduce the following function

$$\phi_{\mathrm{B}}^d(x, y, \mu) = -\frac{f_0(y)}{\mu} + \frac{1}{\mu} \sum_{i=1}^n x_i f_i(y) + \sum_{i=1}^n \ln x_i + n(1 - \ln \mu), \tag{2.59}$$

which is the logarithmic barrier function for $(\mathcal{CD})$, up to a constant factor.

---

[11]On page 101–102 of Fiacco and McCormick's book [35] only a specific example is given for which the dual objective is really increasing.

Lemma 2.19 *We have that* $\phi_B(\bar{y}, \mu) \geq \phi_B^d(x, y, \mu)$ *for all primal feasible* $\bar{y}$ *and dual feasible* (x, y). *Moreover,* $\phi_B(y(\mu), \mu) = \phi_B^d(x(\mu), y(\mu), \mu)$.

Proof: For a convex function $f(y)$ the following inequality is well–known

$$f(\bar{y}) \geq f(y) + \nabla f(y)^T (\bar{y} - y).$$

Hence, since $-f_0(y)$ and $f_i(y)$, $i = 1, \cdots, n$, are convex we may write

$$-f_0(\bar{y}) + f_0(y) + \sum_{i=1}^{n} x_i(f_i(\bar{y}) - f_i(y)) \geq -\nabla f_0(y)^T (\bar{y} - y) +$$

$$\sum_{i=1}^{n} x_i \nabla f_i(y)^T (\bar{y} - y)$$

$$= 0,$$

where the last equality follows from $(\mathcal{CD})$. Using this, we get

$$\phi_B(\bar{y}, \mu) - \phi_B^d(x, y, \mu) = \frac{-f_0(\bar{y}) + f_0(y)}{\mu} - \frac{1}{\mu}\sum_{i=1}^{n} x_i f_i(y) -$$

$$\sum_{i=1}^{n} \ln(-x_i f_i(\bar{y})) - n(1 - \ln \mu)$$

$$\geq -\frac{1}{\mu}\sum_{i=1}^{n} x_i f_i(\bar{y}) - \sum_{i=1}^{n} \ln(-x_i f_i(\bar{y})) -$$

$$n(1 - \ln \mu). \tag{2.60}$$

Now setting the derivatives with respect to x_i of the right–hand side equal to zero, we get

$$x_i = \frac{\mu}{-f_i(\bar{y})}.$$

This choice of x_i minimizes the right–hand side of (2.60), since it is convex in x_i. Substituting this into (2.60), we get

$$\phi_B(\bar{y}, \mu) - \phi_B^d(x, y, \mu) \geq 0.$$

It is easy to verify that equality holds for $\bar{y} = y = y(\mu)$ and $x = x(\mu)$.   □

As a consequence of Lemma 2.19 we have that $(x(\mu), y(\mu))$ maximizes $\phi_B^d(x, y, \mu)$. Now we are ready to give the main result.

Theorem 2.8 *The objective function* $f_0(y(\mu))$ *of the primal problem* $(\mathcal{CP})$ *is monotonically increasing, and the objective function of the dual problem* $(\mathcal{CD})$

$$f_0(y(\mu)) - \sum_{i=1}^{n} x_i(\mu) f_i(y(\mu))$$

is monotonically decreasing if μ *decreases, where* $x(\mu)$ *and* $y(\mu)$ *are defined by (2.5).*

Proof: The first part of the theorem is a classical result of Fiacco and McCormick [35]. (See also the proof of Lemma 2.1.) The proof of the second part of the theorem follows easily by applying Fiacco and McCormick's [35] proof to $\phi_B^d(x, y, \mu)$, since, as a consequence of Lemma 2.19, we have that $(x(\mu), y(\mu))$ maximizes $\phi_B^d(x, y, \mu)$. For completeness we will give the precise proof. Let us denote the dual objective function $f_0(y) - \sum_{i=1}^n x_i f_i(y)$ by $d(x, y)$. Suppose $\bar{\mu} < \mu$. Since $(x(\mu), y(\mu))$ maximizes $\phi_B^d(x, y, \mu)$, and $(x(\bar{\mu}), y(\bar{\mu}))$ maximizes $\phi_B^d(x, y, \bar{\mu})$, we have

$$\phi_B^d(x(\mu), y(\mu), \mu) \geq \phi_B^d(x(\bar{\mu}), y(\bar{\mu}), \mu)$$

and

$$\phi_B^d(x(\bar{\mu}), y(\bar{\mu}), \bar{\mu}) \geq \phi_B^d(x(\mu), y(\mu), \bar{\mu}).$$

These inequalities are equivalent with, respectively,

$$-\frac{d(x(\mu), y(\mu))}{\mu} + \sum_{i=1}^n \ln x_i(\mu) + n(1 - \ln \mu)$$

$$\geq -\frac{d(x(\bar{\mu}), y(\bar{\mu}))}{\mu} + \sum_{i=1}^n \ln x_i(\bar{\mu}) + n(1 - \ln \mu),$$

and

$$-\frac{d(x(\bar{\mu}), y(\bar{\mu}))}{\bar{\mu}} + \sum_{i=1}^n \ln x_i(\bar{\mu}) + n(1 - \ln \bar{\mu})$$

$$\geq -\frac{d(x(\mu), y(\mu))}{\bar{\mu}} + \sum_{i=1}^n \ln x_i(\mu) + n(1 - \ln \bar{\mu}).$$

Adding the two inequalities gives

$$\left(\frac{1}{\bar{\mu}} - \frac{1}{\mu}\right)(d(x(\mu), y(\mu)) - d(x(\bar{\mu}), y(\bar{\mu}))) \geq 0.$$

Since $\bar{\mu} < \mu$ the second part of the lemma follows. $\qquad\square$

2.5.2 The self–concordance condition

In addition to Assumptions 1–3 made in Section 2.1 we assumed an additional smoothness condition in [29], namely that the Hessian matrix of $f_i(y)$, $0 \leq i \leq n$, fulfils the so–called Relative Lipschitz Condition:

$$\exists M > 0: \ \forall v \in \mathbb{R}^m \ \forall y, y + h \in \mathcal{F}^0:$$

$$|v^T(\nabla^2 f_i(y + h) - \nabla^2 f_i(y))v| \leq M\|h\|_H v^T \nabla^2 f_i(y)v. \tag{2.61}$$

This condition was introduced by Jarre [66]. The reader is referred to this paper for a motivation of this condition. It is clear that linear and convex quadratic functions fulfil this condition. In general the condition may be hard to check for a given problem.

In Nesterov and Nemirovsky [119] a different condition is used[12], the so–called self–concordance condition, which is defined as follows:

Definition of self–concordance: A function φ : $\mathcal{F}^0 \to \mathbb{R}$ is called κ–self–concordant on $\mathcal{F}^0$, $\kappa \geq 0$, if φ is three times continuously differentiable in $\mathcal{F}^0$ and for all $y \in \mathcal{F}^0$ and $h \in \mathbb{R}^m$ the following inequality holds:

$$|\nabla^3 \varphi(y)[h, h, h]| \leq 2\kappa \left(h^T \nabla^2 \varphi(y) h \right)^{\frac{3}{2}},$$

where $\nabla^3 \varphi(y)[h, h, h]$ denotes the third differential of φ at y and h.

Note that linear and convex quadratic functions are 0–self–concordant. It is important to observe that this condition relates to the barrier function itself, and not to the individual problem functions $f_i(y)$; i.e. the assumption in [119] is that the barrier function is self–concordant. Jarre [68] proved the following interesting result: if a function f_i fulfils the Relative Lipschitz Condition for infinitesimal $\|h\|$, then the logarithmic barrier function $\varphi(y) := -\ln(-f_i(y))$ is self–concordant with parameter $\kappa = 1 + M$. The converse is not true. So, besides the requirement of three times differentiability, the class of problems for which the logarithmic barrier function fulfils the self–concordance condition is wider than the class of problems which satisfy the Relative Lipschitz Condition.

Since a convex quadratic function $f_i(y)$ fulfils the Relative Lipschitz Condition (with $M = 0$) it follows that $-\ln(-f_i(y))$ is 1–self–concordant. From Lemma A.1 we learn that the sum of 1–self–concordant functions is again a 1–self–concordant function. Hence the result is that the logarithmic barrier function for quadratically constrained convex quadratic programming is 1–self–concordant.

For a general problem it might be hard to check whether its logarithmic barrier function is self–concordant. However, in many cases it is possible to reformulate the problem such that the logarithmic barrier function for the new problem is self–concordant. This is in essence what is done by Nesterov and Nemirovsky in [119], although they describe it as constructing different self–concordant barriers for different problems. Below we will give many classes of problems for which the logarithmic barrier function satisfies the self–concordance condition.

Nesterov and Nemirovsky [119] showed that the logarithmic barrier functions for the following (reformulated) problems are self–concordant:

- linear and convex quadratic programming with convex quadratic constraints;

- primal geometric programming;

- l_p–approximation;

- matrix norm minimization;

- maximal inscribed ellipsoid.

---

[12]Thanks are due to Osman Güler, who explained us monograph [119] in a series of lectures during his stay at Delft in 1991.

In [19] we proved for some other important classes of problems that the logarithmic barrier function (sometimes after a reformulation) satisfies the self–concordance condition. These classes are:

- dual geometric programming;

- extended entropy programming;

- primal l_p–programming;

- dual l_p–programming.

For the precise formulations of these last four problems and for the proofs of self–concordance we refer the reader to Appendix A. In most of these cases the self–concordance parameter appears to be $O(1)$, but in some cases the number of inequality constraints is increased due to the reformulation. (We will see that the complexity bounds depend on both the self–concordance parameter and the number of constraints.)

Besides the Relative Lipschitz Condition and the self–concordance condition, we found two other conditions in the literature, namely the Scaled Lipschitz Condition, introduced by Zhu [170], and a condition used by Monteiro and Adler [110]. These two conditions refer to problems with convex objective function and linear constraints. In Appendix A we describe these conditions more precise, and prove that these conditions are covered by the self–concordance condition if the objective function is three times continuously differentiable. So, again we conclude that self–concordance is the most general condition.

In the remainder of this chapter we will assume that the logarithmic barrier function for $(\mathcal{CP})$

$$\phi_{\mathrm{B}}(y,\mu) = -\frac{f_0(y)}{\mu} - \sum_{i=1}^{n} \ln(-f_i(y))$$

is κ–self-concordant. Without loss of generality we will assume that $\kappa \geq 1$. The strict convexity of $\phi_{\mathrm{B}}(y,\mu)$ follows from the boundedness of $\mathcal{F}$ and the self–concordance of $\phi_{\mathrm{B}}(y,\mu)$ (see [119] and [68]).

2.5.3 Properties near the central path

In this section we deal with some lemmas which will be needed to obtain an upper bound for the total number of outer and inner iterations. Lemmas 2.20 and 2.21 are in essence due to Nesterov and Nemirovsky [119]. As established before, we will denote $H = H(y,\mu)$ if no confusion is possible.

Lemma 2.20 *Let $d \in \mathbb{R}^m$. If $\|d\|_{H(y,\mu)} < \frac{1}{\kappa}$ then $y + d \in \mathcal{F}^0$.*

Proof: In essence, the proof is similar to the proof given in [68], which is a simplified version of the proof in [119]. Since μ is fixed in this lemma we will write $H(y)$ instead of $H(y,\mu)$ in this proof for briefness' sake.

Let $0 \leq t \leq 1$ be such that $y + td \in \mathcal{F}^0$, and $v \in \mathbb{R}^m$ arbitrary. We first show that the norm of v with respect to $H(y)$ and $H(y + td)$ are comparable, i.e.

$$(1 - t\kappa\|d\|_H)\|v\|_H \leq \|v\|_{H(y+td)} \leq \frac{1}{1 - t\kappa\|d\|_H}\|v\|_H. \tag{2.62}$$

To prove these inequalities, we define, for $\rho \in [0, t]$,

$$\Psi(\rho) := \|d\|^2_{H(y+\rho d)} = d^T H(y + \rho d)d$$

and

$$\Lambda(\rho) := \|v\|^2_{H(y+\rho d)} = v^T H(y + \rho d)v.$$

Now we want to evaluate how these norms change if ρ varies. Therefore, we calculate bounds for the derivatives of Ψ and Λ with respect to ρ:

$$\begin{aligned}
|\Psi'(\rho)| &= |\nabla^3\phi_{\mathrm{B}}(y + \rho d, \mu)[d, d, d]| \\
&\leq 2\kappa\|d\|^3_{H(y+\rho d)} \\
&= 2\kappa\Psi(\rho)^{\frac{3}{2}},
\end{aligned} \tag{2.63}$$

and

$$\begin{aligned}
|\Lambda'(\rho)| &= |\nabla^3\phi_{\mathrm{B}}(y + \rho d, \mu)[v, v, d]| \\
&\leq 2\kappa\|v\|^2_{H(y+\rho d)}\|d\|_{H(y+\rho d)} \\
&= 2\kappa\Lambda(\rho)\Psi(\rho)^{\frac{1}{2}},
\end{aligned} \tag{2.64}$$

where the inequalities follow from the fact that ϕ_{B} is κ–self–concordant and from Lemma B.2. From (2.63) we immediately obtain an upper bound for $\Psi(\rho)$

$$\left|\frac{d}{d\rho}\left[\Psi(\rho)^{-\frac{1}{2}}\right]\right| = \left|\frac{1}{2}\Psi(\rho)^{-\frac{3}{2}}\Psi'(\rho)\right| \leq \kappa. \tag{2.65}$$

According to the Mean Value Theorem there exists a $\xi \in (0, \rho)$ such that

$$\Psi(\rho)^{-\frac{1}{2}} = \Psi(0)^{-\frac{1}{2}} + \rho\left.\frac{d}{ds}\left[\Psi(s)^{-\frac{1}{2}}\right]\right|_{s=\xi}.$$

Using the upper bound obtained in (2.65) we get

$$\Psi(\rho)^{-\frac{1}{2}} \geq \Psi(0)^{-\frac{1}{2}} - \rho\kappa = \frac{1}{\|d\|_H} - \rho\kappa,$$

or equivalently

$$\Psi(\rho)^{\frac{1}{2}} \leq \frac{\|d\|_H}{1 - \rho\kappa\|d\|_H}.$$

Substituting this into (2.64) gives

$$|\Lambda'(\rho)| \leq \frac{2\kappa\|d\|_H}{1 - \rho\kappa\|d\|_H}\Lambda(\rho).$$

Now it easily follows that

$$|(\ln \Lambda(\rho))'| = \left|\frac{\Lambda(\rho)'}{\Lambda(\rho)}\right| \leq \frac{2\kappa\|d\|_H}{1 - \rho\kappa\|d\|_H}.$$

Using this we obtain

$$
\begin{aligned}
\left|\ln \frac{\|v\|_{H(y+td)}}{\|v\|_H}\right| &= \frac{1}{2}\left|\ln \frac{\Lambda(t)}{\Lambda(0)}\right| \\
&= \frac{1}{2}|\ln \Lambda(t) - \ln \Lambda(0)| \\
&= \frac{1}{2}\left|\int_0^t (\ln \Lambda(\rho))'d\rho\right| \\
&\leq \int_0^t \frac{\kappa\|d\|_H}{1 - \rho\kappa\|d\|_H}d\rho \\
&= -\ln(1 - \rho\kappa\|d\|_H)|_0^t \\
&= \ln\left(\frac{1}{1 - t\kappa\|d\|_H}\right).
\end{aligned}
$$

Consequently,

$$\frac{\|v\|_{H(y+td)}}{\|v\|_H} \leq \frac{1}{1 - t\kappa\|d\|_H}$$

and

$$\frac{\|v\|_H}{\|v\|_{H(y+td)}} \leq \frac{1}{1 - t\kappa\|d\|_H},$$

which proves (2.62). Since $\|d\|_H < \frac{1}{\kappa}$, we have from (2.62) that $H(y+td)$ is bounded for all $0 \leq t \leq 1$, and thus $\phi_B(y + td, \mu)$ is bounded. On the other hand, ϕ_B takes infinite values on the boundary of the feasible set. Consequently, $y + d \in \mathcal{F}^0$. $\square$

The following lemma, proved in [119], gives a quadratic convergence result. Recall that $p = p(y, \mu)$ is the Newton direction in y.

Lemma 2.21 Let $y^+ := y + p$. If $\|p\|_H < \frac{1}{\kappa}$ then $y^+ \in \mathcal{F}^0$, and

$$\|p(y^+, \mu)\|_{H(y^+, \mu)} \leq \frac{\kappa}{(1 - \kappa\|p\|_H)^2}\|p\|_H^2.$$

Proof: Again, the proof is similar to the proof given in [68], which is a simplified version of the proof in [119]. Since μ is fixed in this lemma, we will write, for briefness' sake, $g(y)$, $H(y)$ and $p(y)$ instead of $g(y,\mu)$, $H(y,\mu)$ and $p(y,\mu)$ in this proof.

Since $\|p\|_H < \frac{1}{\kappa}$, we have due to Lemma 2.20 that $y + tp \in \mathcal{F}^0$, for all $0 \leq t \leq 1$, and according to (2.62)

$$|v^T (H(y) - H(y + tp))v| \leq \left(\frac{1}{(1 - t\kappa\|p\|_H)^2} - 1\right)\|v\|_H^2, \qquad (2.66)$$

for arbitrary $v \in \mathbb{R}^m$. Using the generalized Cauchy–Schwarz inequality of Lemma B.1, we have

$$
\begin{aligned}
\left| \frac{d}{dt} g(y+tp)^T v - p^T H(y) v \right| &= \left| p^T \left(H(y+tp) - H(y) \right) v \right| \\
&\leq \sqrt{\left| p^T \left(H(y+tp) - H(y) \right) p \right|} \times \\
&\qquad \sqrt{\left| v^T \left(H(y+tp) - H(y) \right) v \right|} \\
&\leq \left(\frac{1}{(1 - t\kappa \|p\|_H)^2} - 1 \right) \|v\|_H \|p\|_H, \quad (2.67)
\end{aligned}
$$

where the last inequality follows from (2.66). Note that the left–hand side of (2.67) is the absolute value of the derivative $\varphi'(t)$, where

$$
\varphi(t) := g(y+tp)^T v - (1-t) g(y)^T v, \qquad (2.68)
$$

since $p = -H^{-1} g$. By integration we obtain an upper bound for $|\varphi(t)|$:

$$
\begin{aligned}
|\varphi(t)| &= \left| \int_0^t \varphi'(s) ds \right| \\
&\leq \|v\|_H \|p\|_H \int_0^t \left(\frac{1}{(1 - s\kappa \|p\|_H)^2} - 1 \right) ds \\
&= \kappa \|v\|_H \|p\|_H^2 \frac{t^2}{1 - t\kappa \|p\|_H}, \qquad (2.69)
\end{aligned}
$$

where the inequality follows from (2.67) and (2.68). For $t = 1$, $y^+ = y + p$, this implies

$$
|\varphi(1)| = |g(y^+)^T v| \leq \kappa \|v\|_H \|p\|_H^2 \frac{1}{1 - \kappa \|p\|_H}. \qquad (2.70)
$$

Now choosing $v = p(y^+) = -H(y^+)^{-1} g(y^+)$ we finally obtain

$$
\begin{aligned}
\|p(y^+)\|_{H(y^+)}^2 &= |g(y^+)^T p(y^+)| \\
&\leq \frac{\kappa \|p(y^+)\|_H \|p\|_H^2}{1 - \kappa \|p\|_H} \\
&\leq \frac{\kappa \|p\|_H^2 \|p(y^+)\|_{H(y^+)}}{(1 - \kappa \|p\|_H)^2},
\end{aligned}
$$

where the first inequality follows from (2.70) and the last inequality from (2.62). This proves the lemma. $\qquad \square$

For $\|p\|_H < \frac{3 - \sqrt{5}}{2\kappa}$, the result of Lemma 2.21 is that $\|p(y^+, \mu)\|_{H(y^+, \mu)} < \|p\|_H$, and hence convergence of Newton's method. For $\|p\|_H \leq \frac{1}{3\kappa}$ the lemma gives

$$
\|p(y^+, \mu)\|_{H(y^+, \mu)} \leq \frac{9}{4} \kappa \|p\|_H^2. \qquad (2.71)
$$

Note that for linear and quadratic programming ($\kappa = 1$) the result of Lemma 2.21 reduces to

$$\|p(y^+, \mu)\|_{H(y^+,\mu)} \leq \frac{1}{(1 - \|p\|_H)^2}\|p\|_H^2,$$

which is worse than the pure quadratic convergence result

$$\|p(y^+, \mu)\|_{H(y^+,\mu)} \leq \|p\|_H^2,$$

obtained in Lemmas 2.3 and 2.12.

The following lemma gives an upper bound for the difference in the barrier function value of an approximately centered iterate and the exact center.

Lemma 2.22 *If* $\|p\|_H \leq \frac{1}{3\kappa}$ *then*

$$\phi_B(y, \mu) - \phi_B(y(\mu), \mu) \leq \frac{\|p\|_H^2}{1 - \left(\frac{9}{4}\kappa\|p\|_H\right)^2}.$$

Proof: The barrier function ϕ_B is convex in y, whence

$$\phi_B(y + p, \mu) \geq \phi_B(y, \mu) + p^T g.$$

Now using that $p = -H^{-1}g$, we have

$$p^T g = -p^T H p = -\|p\|_H^2. \tag{2.72}$$

Substitution gives

$$\phi_B(y, \mu) - \phi_B(y + p, \mu) \leq \|p\|_H^2. \tag{2.73}$$

Now let $y^0 := y$ and let $y^1, y^2, \ldots$ denote the sequence of points obtained by repeating Newton steps, starting at y^0. By (2.71) we have

$$\begin{aligned}
\|p(y^i, \mu)\|_{H(y^i,\mu)} &\leq \left(\frac{9}{4}\kappa\right)^{2^i-1} \|p(y^0, \mu)\|_{H(y^0,\mu)}^{2^i} \\
&= \left(\frac{9}{4}\kappa\right)^{2^i-1} \|p\|_H^{2^i}. \tag{2.74}
\end{aligned}$$

This means that $\|p(y^i, \mu)\|_{H(y^i,\mu)} \to 0$ if $i \to \infty$. Since all y^i lie in a bounded region and $\delta(y, \mu) = 0$ if and only if $y = y(\mu)$, we have that all limit points of the sequence are $y(\mu)$. Hence, the sequence of Newton iterates converges to $y(\mu)$. So, using (2.73) and (2.74), we may write

$$\begin{aligned}
\phi_B(y, \mu) - \phi_B(y(\mu), \mu) &= \sum_{i=0}^{\infty} \left(\phi_B(y^i, \mu) - \phi_B(y^{i+1}, \mu)\right) \\
&\leq \sum_{i=0}^{\infty} \|p(y^i, \mu)\|_{H(y^i,\mu)}^2 \\
&\leq \sum_{i=0}^{\infty} \left(\frac{9}{4}\kappa\right)^{2^{i+1}-2} \|p\|_H^{2^{i+1}} \\
&\leq \frac{\|p\|_H^2}{1 - \left(\frac{9}{4}\kappa\|p\|_H\right)^2}.
\end{aligned}$$

□

The following lemma gives an upper bound for the difference of the objective value in the exact center and an approximately centered iterate.

Lemma 2.23 *If* $\|p\|_H \le \frac{1}{3\kappa}$, *then*

$$|f_0(y) - f_0(y(\mu))| \le \frac{\|p\|_H}{1 - \frac{9}{4}\kappa\|p\|_H} \frac{1 + \kappa\|p\|_H^2}{1 - \kappa\|p\|_H} \mu\sqrt{n}.$$

Proof: Since $-f_0(y)$ is convex, we have

$$\nabla f_0(y + p)^T p \le f_0(y + p) - f_0(y) \le \nabla f_0(y)^T p. \qquad (2.75)$$

We first derive an upper bound for the right–hand side expression. From (2.72) we have $p^T g = -\|p\|_H^2$. On the other hand from (2.2) it follows that

$$-p^T g = p^T \left(\frac{\nabla f_0(y)}{\mu} - \sum_{i=1}^{n} \frac{\nabla f_i(y)}{-f_i(y)} \right). \qquad (2.76)$$

So we have

$$\frac{\nabla f_0(y)^T p}{\mu} = \|p\|_H^2 + p^T \sum_{i=1}^{n} \frac{\nabla f_i(y)}{-f_i(y)}. \qquad (2.77)$$

Now let J denote the $m \times n$ matrix whose columns are $\frac{\nabla f_i(y)}{-f_i(y)}$. It is easy to see that

$$\begin{aligned}
\left| p^T \sum_{i=1}^{n} \frac{\nabla f_i(y)}{-f_i(y)} \right| &= |p^T J e| \\
&\le \|J^T p\|\|e\| \\
&\le \sqrt{n}\|p\|_H, \qquad (2.78)
\end{aligned}$$

where the last inequality follows since

$$\begin{aligned}
\|J^T p\|^2 &= p^T J J^T p \\
&\le p^T H p \\
&= \|p\|_H^2.
\end{aligned}$$

(Note that $J J^T = \sum_{i=1}^{m} \nabla f_i(y) \nabla f_i(y)^T / f_i(y)^2$ and that $\sum_{i=1}^{n} \nabla f_i(y) \nabla f_i(y)^T / f_i(y)^2 \preceq H$, see (2.3).) Substituting (2.78) into (2.77) gives

$$\nabla f_0(y)^T p \le \mu \left(\|p\|_H^2 + \sqrt{n}\|p\|_H \right). \qquad (2.79)$$

Now we derive a lower bound for the left–hand side expression in (2.75). Similar as in (2.77) we have

$$\frac{\nabla f_0(y + p)^T p}{\mu} = -p^T g(y + p) + p^T \sum_{i=1}^{n} \frac{\nabla f_i(y + p)}{-f_i(y + p)}. \qquad (2.80)$$

From (2.70), with $v = p$, it follows that

$$|p^T g(y + p)| \leq \frac{\kappa \|p\|_H^3}{1 - \kappa \|p\|_H}.$$

From (2.62) and using similar arguments as in (2.78) it follows that

$$\left| p^T \sum_{i=1}^n \frac{\nabla f_i(y + p)}{-f_i(y + p)} \right| \leq \sqrt{n} \|p\|_{H(y+p,\mu)}$$

$$\leq -\sqrt{n} \frac{\|p\|_H}{1 - \kappa \|p\|_H}.$$

Substituting the last two formulas into (2.80) yields

$$\nabla f_0(y + p)^T p \geq -\mu \|p\|_H \frac{\sqrt{n} + \kappa \|p\|_H^2}{1 - \kappa \|p\|_H}. \tag{2.81}$$

Combining (2.75), (2.79) and (2.81) gives

$$|f_0(y + p) - f_0(y)| \leq \mu \sqrt{n} \|p\|_H \frac{1 + \kappa \|p\|_H^2}{1 - \kappa \|p\|_H}. \tag{2.82}$$

Again, let $y^0 := y$ and let $y^1, y^2, \cdots$ denote the sequence of points obtained by repeating Newton steps, starting at y^0. Then we have

$$|f_0(y(\mu)) - f_0(y)| = \left| \sum_{i=0}^{\infty} \left(f_0(y^{i+1}) - f_0(y^i) \right) \right|$$

$$\leq \sum_{i=0}^{\infty} \left| f_0(y^{i+1}) - f_0(y^i) \right|$$

$$\leq \sum_{i=0}^{\infty} \|p(y^i, \mu)\|_{H(y^i,\mu)} \left(\frac{1 + \kappa \|p(y^i,\mu)\|_{H(y^i,\mu)}^2}{1 - \kappa \|p(y^i,\mu)\|_{H(y^i,\mu)}} \right) \cdot \mu \sqrt{n}$$

$$\leq \frac{1 + \kappa \|p\|_H^2}{1 - \kappa \|p\|_H} \mu \sqrt{n} \sum_{i=0}^{\infty} \|p(y^i, \mu)\|_{H(y^i,\mu)}$$

$$\leq \frac{1 + \kappa \|p\|_H^2}{1 - \kappa \|p\|_H} \mu \sqrt{n} \sum_{i=0}^{\infty} \left(\frac{9}{4} \kappa \right)^{2^i - 1} \|p\|_H^{2^i}$$

$$\leq \frac{1 + \kappa \|p\|_H^2}{1 - \kappa \|p\|_H} \frac{\|p\|_H}{1 - \frac{9}{4} \kappa \|p\|_H} \mu \sqrt{n},$$

where the second inequality follows from (2.82) and the fourth inequality from (2.74).
□

2.5.4 Complexity analysis

Based on the lemmas in the previous section, we will give upper bounds for the total number of outer and inner iterations for the Logarithmic Barrier Algorithm applied to $(\mathcal{CP})$. First, the complexity analysis is carried out for the long– and medium–step variant, and finally for the short–step variant. We set $\tau = \frac{1}{3\kappa}$.

Theorem 2.9 *After at most*

$$\frac{1}{\theta} \ln \frac{4n\mu_0}{\epsilon}$$

outer iterations, the Logarithmic Barrier Algorithm ends up with a solution for $(\mathcal{CP})$
such that $z^* - f_0(y) \leq \epsilon$.

Proof: The proof is similar as the proof of Theorem 2.2. For completeness we will give the proof in detail. The algorithm stops when $\mu_K = (1 - \theta)^K \mu_0 \leq \frac{\epsilon}{4n}$. Taking logarithms we require

$$-K \ln(1 - \theta) \geq \ln \frac{4n\mu_0}{\epsilon}.$$

Since $\theta \leq -\ln(1 - \theta)$, this certainly holds if

$$K \geq \frac{1}{\theta} \ln \frac{4n\mu_0}{\epsilon}.$$

From (2.6) we derive that

$$z^* - f_0(y(\mu_K)) \leq n\mu_K,$$

and from Lemma 2.23, with $\|p\| \leq \frac{1}{3\kappa}$,

$$f_0(y(\mu_K)) - f_0(y) \leq \frac{\frac{1}{3\kappa}}{1 - \frac{9}{4}\kappa.\frac{1}{3\kappa}} \frac{1 + \kappa(\frac{1}{3\kappa})^2}{1 - \kappa.\frac{1}{3\kappa}} \mu\sqrt{n}$$

$$\leq 3\sqrt{n}\mu.$$

Using these upper bounds, we can derive an upper bound for the gap $z^* - f_0(y)$ after the algorithm has stopped

$$\begin{aligned}
z^* - f_0(y) &= z^* - f_0(y(\mu_K)) + f_0(y(\mu_K)) - f_0(y) \\
&\leq \mu_K(n + 3\sqrt{n}) \\
&\leq \frac{\epsilon}{4n}(n + 3\sqrt{n}) \\
&\leq \epsilon.
\end{aligned}$$

□

The following lemma states that if we do a line search along the Newton direction, then a sufficient decrease in the logarithmic barrier function value can be guaranteed.

Lemma 2.24 *Let* $\bar{\alpha} := \frac{1}{1 + \kappa\|p\|_H}$. *Then*

$$\phi_B(y, \mu) - \phi_B(y + \bar{\alpha}p, \mu) \geq \frac{1}{\kappa^2}\left(\kappa\|p\|_H - \ln(1 + \kappa\|p\|_H)\right).$$

Proof: The proof is a modified version of the proof given in [119]. Let $\Delta\phi_{\text{B}}(\alpha) := \phi_{\text{B}}(y,\mu) - \phi_{\text{B}}(y + \alpha p, \mu)$. From (2.68), (2.69), with $v = p$, and (2.72) we derive

$$
\begin{aligned}
\frac{d}{d\,\alpha}\Delta\phi_{\text{B}}(\alpha) &= -g(y + \alpha p, \mu)^T p \\
&= -\varphi(\alpha) - (1 - \alpha)g(y)^T p \\
&= -\varphi(\alpha) + (1 - \alpha)\|p\|_H^2 \\
&\geq (1 - \alpha)\|p\|_H^2 - \kappa\|p\|_H^3 \frac{\alpha^2}{1 - \alpha\kappa\|p\|_H}.
\end{aligned}
$$

Consequently

$$
\begin{aligned}
\Delta\phi_{\text{B}}(\alpha) &= \Delta\phi_{\text{B}}(\alpha) - \Delta\phi_{\text{B}}(0) \\
&= \int_0^\alpha \frac{d}{d\,s}\Delta\phi_{\text{B}}(s)\,d\,s \\
&\geq \int_0^\alpha \left((1 - s)\|p\|_H^2 - \kappa\|p\|_H^3 \frac{s^2}{1 - s\kappa\|p\|_H}\right) d\,s \\
&= \frac{1}{\kappa^2}\left(\kappa^2\|p\|_H^2\alpha + \kappa\|p\|_H\alpha + \ln(1 - \alpha\kappa\|p\|_H)\right). \quad (2.83)
\end{aligned}
$$

The right–hand side is maximal for $\bar{\alpha} = \frac{1}{1+\kappa\|p\|_H}$. Substituting this value into (2.83) yields the lemma. $\quad\square$

Note the similarity between this lemma and Lemmas 2.6 and 2.15.

Theorem 2.10 *Each outer iteration requires at most*

$$
\frac{22\theta}{(1 - \theta)^2}\left(\theta\kappa^2 n + \frac{5}{2}\kappa\sqrt{n}\right) + \frac{22}{3}
$$

inner iterations.

Proof: We denote the barrier parameter value in an arbitrary outer iteration by $\bar{\bar{\mu}}$, while the parameter value in the previous outer iteration is denoted by $\bar{\mu}$. The iterate at the beginning of the outer iteration is denoted by y. Hence y is centered with respect to $y(\bar{\mu})$ and $\bar{\bar{\mu}} = (1 - \theta)\bar{\mu}$. Note that because of Lemma 2.24 during each inner iteration the decrease in the barrier function value is at least

$$
\frac{1}{\kappa^2}(\kappa\|p\|_H - \ln(1 + \kappa\|p\|_H)).
$$

Since this function is increasing in $\|p\|_H$ and during each inner iteration we have $\|p\|_H \geq \frac{1}{3\kappa}$ it follows that the decrease is at least

$$
\frac{1}{\kappa^2}\left(\frac{1}{3} - \ln(1 + \frac{1}{3})\right) > \frac{1}{22\kappa^2}.
$$

Now let N denote the number of inner iterations during one outer iteration. Since the gap between the barrier function value in the current iterate y and the next center $y(\bar{\bar{\mu}})$ is equal to $\phi_{\mathrm{B}}(y, \bar{\bar{\mu}}) - \phi_{\mathrm{B}}(y(\bar{\bar{\mu}}), \bar{\bar{\mu}})$, we have

$$\frac{N}{22\kappa^2} \leq \phi_{\mathrm{B}}(y, \bar{\bar{\mu}}) - \phi_{\mathrm{B}}(y(\bar{\bar{\mu}}), \bar{\bar{\mu}}). \tag{2.84}$$

Let us call the right–hand side of (2.84) $\Phi_{\mathrm{B}}(y, \bar{\bar{\mu}})$, i.e.

$$\Phi_{\mathrm{B}}(y, \bar{\bar{\mu}}) = \phi_{\mathrm{B}}(y, \bar{\bar{\mu}}) - \phi_{\mathrm{B}}(y(\bar{\bar{\mu}}), \bar{\bar{\mu}}).$$

According to the Mean Value Theorem there is a $\hat{\mu} \in (\bar{\bar{\mu}}, \bar{\mu})$ such that

$$\Phi_{\mathrm{B}}(y, \bar{\bar{\mu}}) = \Phi_{\mathrm{B}}(y, \bar{\mu}) + \left. \frac{d\,\Phi_{\mathrm{B}}(y, \mu)}{d\,\mu} \right|_{\mu = \hat{\mu}} (\bar{\bar{\mu}} - \bar{\mu}). \tag{2.85}$$

Let us now look at $\frac{d\,\Phi_{\mathrm{B}}(y,\mu)}{d\,\mu}$. We have

$$\frac{d\,\phi_{\mathrm{B}}(y, \mu)}{d\,\mu} = \frac{f_0(y)}{\mu^2},$$

and, denoting the derivative of $y(\mu)$ with respect to μ by y',

$$\begin{aligned}
\frac{d\,\phi_{\mathrm{B}}(y(\mu), \mu)}{d\,\mu} &= \frac{f_0(y(\mu))}{\mu^2} - \frac{\nabla f_0(y(\mu))^T y'}{\mu} + \sum_{i=1}^n \frac{\nabla f_i(y(\mu))^T y'}{-f_i(y(\mu))} \\
&= \frac{f_0(y(\mu))}{\mu^2} - \frac{\nabla f_0(y(\mu))^T y'}{\mu} + \frac{1}{\mu} \sum_{i=1}^n x_i \nabla f_i(y(\mu))^T y' \\
&= \frac{f_0(y(\mu))}{\mu^2},
\end{aligned}$$

where the last two equations follow from (2.5). So

$$-\left. \frac{d\,\Phi_{\mathrm{B}}(y, \mu)}{d\,\mu} \right|_{\mu = \hat{\mu}} = \left. \frac{f_0(y(\mu)) - f_0(y)}{\mu^2} \right|_{\mu = \hat{\mu}} \leq \frac{|f_0(y(\bar{\bar{\mu}})) - f_0(y)|}{\bar{\bar{\mu}}^2},$$

where the last inequality follows from the fact that $\bar{\bar{\mu}} < \hat{\mu}$ and from the monotonicity property of the primal objective (Theorem 2.8). Substituting this into (2.85) gives

$$\begin{aligned}
\Phi_{\mathrm{B}}(y, \bar{\bar{\mu}}) &\leq \Phi_{\mathrm{B}}(y, \bar{\mu}) + \frac{|f_0(y(\bar{\bar{\mu}})) - f_0(y)|}{\bar{\bar{\mu}}^2}(\bar{\mu} - \bar{\bar{\mu}}) \\
&\leq \Phi_{\mathrm{B}}(y, \bar{\mu}) + \\
&\quad \left(\frac{|f_0(y(\bar{\mu})) - f_0(y)|}{\bar{\bar{\mu}}} + \frac{f_0(y(\bar{\bar{\mu}})) - f_0(y(\bar{\mu}))}{\bar{\bar{\mu}}} \right) \frac{(\bar{\mu} - \bar{\bar{\mu}})}{\bar{\bar{\mu}}}. \tag{2.86}
\end{aligned}$$

Since $\|p(y, \bar{\mu})\|_{H(y,\bar{\mu})} \leq \frac{1}{3\kappa}$, we have due to Lemma 2.22

$$\Phi_{\mathrm{B}}(y, \bar{\mu}) \leq \frac{1}{3\kappa^2}.$$

Now note that due to Lemma 2.23 and $\|p(y, \bar{\mu})\|_{H(y, \bar{\mu})} \leq \frac{1}{3\kappa}$,

$$|f_0(y(\bar{\mu})) - f_0(y)| \leq \frac{5}{2} \frac{\bar{\mu}\sqrt{n}}{\kappa}.$$

Moreover, because of Lemma 2.8, we have

$$
\begin{aligned}
f_0(y(\bar{\bar{\mu}})) - f_0(y(\bar{\mu})) &\leq \sum_{i=1}^{n} x_i(\bar{\mu}) f_i(y(\bar{\bar{\mu}})) - \sum_{i=1}^{n} x_i(\bar{\mu}) f_i(y(\bar{\mu})) \\
&= n(\bar{\mu} - \bar{\bar{\mu}}) \\
&= \theta n\bar{\mu}.
\end{aligned}
$$

Plugging all these upper bounds in (2.86) gives

$$
\begin{aligned}
\Phi_{\text{B}}(y, \bar{\bar{\mu}}) &\leq \frac{1}{3\kappa^2} + \left(\frac{5\sqrt{n}}{2\kappa(1-\theta)} + \frac{\theta n}{1-\theta}\right) \frac{\theta}{1-\theta} \\
&= \frac{1}{3\kappa^2} + \frac{\theta}{(1-\theta)^2}\left(\frac{5\sqrt{n}}{2\kappa} + \theta n\right).
\end{aligned}
$$

Substituting this into (2.84) it follows that

$$\frac{N}{22\kappa^2} \leq \frac{1}{3\kappa^2} + \frac{\theta}{(1-\theta)^2}\left(\frac{5}{2\kappa}\sqrt{n} + \theta n\right),$$

from which the theorem follows.   □

Combining Theorems 2.9 and 2.10, the total number of iterations turns out to be given by the following theorem.

Theorem 2.11 *An upper bound for the total number of Newton iterations is given by*

$$\left[\frac{22}{(1-\theta)^2}\left(\theta\kappa^2 n + \frac{5}{2}\kappa\sqrt{n}\right) + \frac{22}{3\theta}\right]\ln\frac{4n\mu_0}{\epsilon}.$$

                                                                    □

This makes clear that to obtain an ϵ-optimal solution the algorithm needs

- $O(\kappa^2 n \ln \frac{n\mu_0}{\epsilon})$ Newton iterations for the long–step variant $(0 < \theta < 1)$;

- $O(\kappa^2 \sqrt{n} \ln \frac{n\mu_0}{\epsilon})$ Newton iterations for the medium–step variant $(\theta = \frac{\nu}{\sqrt{n}}, \nu > 0)$;

- $O(\kappa\sqrt{n} \ln \frac{n\mu_0}{\epsilon})$ Newton iterations for $\theta = \frac{\nu}{\kappa\sqrt{n}}, \nu > 0$.

Note that the last variant, which is a special case of the medium–step variant, yields the best complexity result.

We want to point out that a complexity analysis for the short–step path–following method can easily be given by using some of the lemmas given above. Short–step path–following methods start at a nearly centered iterate and after the parameter is reduced by a small factor, a unit Newton step is taken. The reduction parameter is sufficiently small, such that the new iterate is again nearly centered with respect to the new center.

Lemma 2.25 is needed for analyzing such a short–step path–following method. It shows how the distance changes if the barrier parameter μ is reduced.

Lemma 2.25 *Let* $\mu^+ := (1 - \theta)\mu$. *Then*

$$\|p(y,\mu^+)\|_{H(y,\mu^+)} \le \frac{1}{1-\theta}\left(\|p\|_H + \theta\sqrt{n}\right).$$

Proof: Let us first introduce the notation g^+, H^+ and p^+ for $g(y,\mu^+)$, $H(y,\mu^+)$ and $p(y,\mu^+)$, respectively. It follows from (2.3) that

$$H^+ = H - \frac{\theta}{\mu^+}\nabla^2 f_0(y).$$

Since $-f_0(y)$ is convex, it follows from Lemma B.3 that

$$(H^+)^{-1} \preceq H^{-1}.$$

Using this result, we obtain for the distance after reducing μ the following upper bound:

$$
\begin{aligned}
\|p^+\|_{H^+} &= \sqrt{(g^+)^T(H^+)^{-1}g^+}\\
&\le \sqrt{(g^+)^T H^{-1}g^+}\\
&= \|g^+\|_{H^{-1}}\\
&= \left\|-\frac{\nabla f_0(y)}{\mu^+} + \sum_{i=1}^n \frac{\nabla f_i(y)}{-f_i(y)}\right\|_{H^{-1}}\\
&= \left\|\frac{1}{1-\theta}\left(-\frac{\nabla f_0(y)}{\mu} + \sum_{i=1}^n \frac{\nabla f_i(y)}{-f_i(y)}\right) - \frac{\theta}{1-\theta}\sum_{i=1}^n \frac{\nabla f_i(y)}{-f_i(y)}\right\|_{H^{-1}}\\
&\le \frac{1}{1-\theta}\left(\left\|-\frac{\nabla f_0(y)}{\mu} + \sum_{i=1}^n \frac{\nabla f_i(y)}{-f_i(y)}\right\|_{H^{-1}} + \theta\left\|\sum_{i=1}^n \frac{\nabla f_i(y)}{-f_i(y)}\right\|_{H^{-1}}\right)\\
&= \frac{1}{1-\theta}\left(\|p\|_H + \theta\left\|\sum_{i=1}^n \frac{\nabla f_i(y)}{-f_i(y)}\right\|_{H^{-1}}\right).
\end{aligned}
\tag{2.87}
$$

Let us now continue by evaluating $\| \sum_{i=1}^{n} \frac{\nabla f_i(y)}{-f_i(y)} \|_{H^{-1}}$:

$$
\left\| \sum_{i=1}^{n} \frac{\nabla f_i(y)}{-f_i(y)} \right\|_{H^{-1}}^{2} = \left(\sum_{i=1}^{n} \frac{\nabla f_i(y)}{-f_i(y)} \right)^{T} \left(-\frac{\nabla^2 f_0(y)}{\mu} + \sum_{i=1}^{n} \frac{\nabla^2 f_i(y)}{-f_i(y)} + \right.
$$

$$
\left. \sum_{i=1}^{n} \frac{\nabla f_i(y) \nabla f_i(y)^{T}}{f_i(y)^2} \right)^{-1} \sum_{i=1}^{n} \frac{\nabla f_i(y)}{-f_i(y)}
$$

$$
= e^{T} J^{T} \left(-\frac{\nabla^2 f_0(y)}{\mu} + \sum_{i=1}^{n} \frac{\nabla^2 f_i(y)}{-f_i(y)} + JJ^{T} \right)^{-1} Je, \quad (2.88)
$$

where J is again defined as the $m \times n$ matrix whose columns are $\frac{\nabla f_i(y)}{-f_i(y)}$. From Lemma B.3 it follows that the eigenvalues of

$$
J^{T} \left(-\frac{\nabla^2 f_0(y)}{\mu} + \sum_{i=1}^{n} \frac{\nabla^2 f_i(y)}{-f_i(y)} + JJ^{T} \right)^{-1} J
$$

are all smaller than or equal to one. Consequently from (2.88) we get

$$
\left\| \sum_{i=1}^{n} \frac{\nabla f_i(y)}{-f_i(y)} \right\|_{H^{-1}}^{2} \leq n.
$$

Substituting this into (2.87) gives

$$
\|p^{+}\|_{H^{+}} \leq \frac{1}{1 - \theta} \left(\|p\|_{H} + \theta \sqrt{n} \right),
$$

which proves the lemma. $\qquad \square$

Theorem 2.12 *Let* $y^{+} := y + p$ *and* $\mu^{+} := (1 - \theta)\mu$, *where* $\theta = \frac{1}{30\kappa\sqrt{n}}$. *If* $\|p\|_H \leq \frac{1}{3\kappa}$ *then*

$$
\|p(y^{+}, \mu^{+})\|_{H(y^{+}, \mu^{+})} \leq \frac{1}{3\kappa}.
$$

Proof: Using Lemma 2.21 and Lemma 2.25, we have

$$
\|p(y^{+}, \mu^{+})\|_{H(y^{+}, \mu^{+})} \leq \frac{9}{4} \kappa \|p(y, \mu^{+})\|_{H(y, \mu^{+})}^{2}
$$

$$
\leq \frac{9}{4} \kappa \left[\frac{1}{1 - \frac{1}{30\kappa\sqrt{n}}} \left(\frac{1}{3\kappa} + \frac{1}{30\kappa} \right) \right]^{2}
$$

$$
< \frac{1}{3\kappa},
$$

which proves the theorem. $\qquad \square$

Consequently, according to Theorem 2.12, if θ is small, then one unit Newton step is sufficient to reach the vicinity of $y(\mu^{+})$. With the help of Theorem 2.9 it is easy to see that this short–step algorithm requires $O(\kappa\sqrt{n} \ln \frac{n\mu_0}{\epsilon})$ Newton iterations.

2.6   Miscellaneous remarks

Further remarks on the logarithmic barrier method

In each iteration of the logarithmic barrier method one has to calculate $(AS^{-2}A^T)^{-1}$ (for linear programming) or $(AH^{-1}A^T)^{-1}$ (for quadratic programming), which costs $O(n^3)$ arithmetic operations. Consequently, the overall complexity bound (counted in arithmetic operations) is $O(n^4 \ln \frac{n\mu_0}{\epsilon})$ for the long–step variant and $O(n^3 \sqrt{n} \ln \frac{n\mu_0}{\epsilon})$ for the medium– and short–step variant. In Section 4.1 we will show that for linear programming these figures can be brought down to $O(n^3 \sqrt{n} \ln \frac{n\mu_0}{\epsilon})$ and $O(n^3 \ln \frac{n\mu_0}{\epsilon})$ respectively, by using approximate solutions and rank–one updates.

Even though our analysis for quadratic programming is based on the null–space form (2.42) for the Newton direction, in practice either the null–space or the range–space form (2.41) can be used. It is obvious that the null–space form is more efficient when the number of linear constraints is relatively large compared to the number of variables. The range–space form is efficient when the number of linear constraints is small compared to the number of variables. In our analysis we made the assumption that $X^{-1}Z$ is orthonormal. However, the search direction does not change if Z is any basis for the null–space of A. So, in practice we do not have to do all the work to find an orthonormal Z in each iteration. We refer the reader to Gill et al. [41] for the numerical aspects.

Note that there is a discrepancy in the complexity bounds derived in the previous sections. We would expect the complexity bounds for the long–step version to be better than for the medium– and short–step version. However, the contrary is true! One reason for this might be that the derived number of updates in the barrier parameter is exact, while the upper bound for the number of iterations needed to return to the vicinity of the central trajectory can be very pessimistic, since

- this upper bound is based on fixed stepsizes, while it is allowed to do line searches in our algorithm

- in many inner iterations we will have $\|p\|_H \gg \tau$, giving larger reductions in the barrier function value than used in the analysis.

The short–step logarithmic barrier method can be speeded up by solving

$$\min\{\mu \ : \ \|p(y,\mu)\|_{H(y,\mu)} \le \sqrt{\tau}\}, \tag{2.89}$$

instead of simply reducing μ by a fixed factor $1 - \theta$. Note that if this updating technique is used, the resulting complexity is at least as good as for the fixed updating scheme! For linear programming, (2.89) turns out to be equivalent to solving a quadratic equation. For general convex programming, (2.89) is more complicated, but if we reformulate the problem such that the objective is linear (which is always possible), it is again equivalent to solving a quadratic equation.

The results obtained in Section 2.4 for convex quadratic programming are the same as the corresponding results obtained in Section 2.3 for linear programming. Moreover, since for linear and convex quadratic programming problems $\kappa = 1$, we have that the complexity results obtained in Section 2.5 for these problems are the same as those obtained in Sections 2.3 and 2.4.

Linear equalities in $(\mathcal{CP})$

Note that in the general convex programming problem $(\mathcal{CP})$ there are only inequalities. However, a mixture of inequalities $(f_i(y) \leq 0)$ and linear equalities $(Gy = d)$ can also be handled by the method described in this chapter.

One way to deal with such problems is to split the matrix G into a basis B and a rest matrix R, and then eliminate the corresponding basis variables:

$$y_B = B^{-1}d - B^{-1}R y_R.$$

An other way is to use projected Newton directions instead of pure Newton directions. The Newton search direction can then be obtained by solving the following equations (cf. (2.39))

$$g + Hp = A^T y, \quad Ap = 0.$$

An explicit expression for p is given by (2.41):

$$p = -H^{-\frac{1}{2}} P_{AH^{-\frac{1}{2}}} H^{-\frac{1}{2}} g.$$

Relaxing the initial centering condition

The initial centering condition $\|p(y^0, \mu_0)\|_{H(y^0, \mu_0)} \leq \tau$ can be relaxed to

$$\phi_B(y^0, \mu_0) - \phi_B(y(\mu_0), \mu_0) \leq O\left(\sqrt{n} \ln \frac{n\mu_0}{\epsilon}\right)$$

for the medium–step version, and to

$$\phi_B(y^0, \mu_0) - \phi_B(y(\mu_0), \mu_0) \leq O\left(n \ln \frac{n\mu_0}{\epsilon}\right)$$

for the long–step version. This holds because Lemma 2.24 states that in each inner iteration the logarithmic barrier function value decreases at least with $\frac{1}{22\kappa^2}$. So, starting from the initial point y^0 we find a point close to $y(\mu_0)$ in $O\left(\kappa^2 \sqrt{n} \ln \frac{n\mu_0}{\epsilon}\right)$ and $O\left(\kappa^2 n \ln \frac{n\mu_0}{\epsilon}\right)$ inner Newton iterations, respectively. The same order of Newton iterations is necessary to obtain an ϵ–optimal solution from this point. Not that for the long–step algorithm for $(\mathcal{LD})$ with a bounded feasible region $\mathcal{F}$, with $\epsilon = 2^{-2L}$ and $\mu_0 \leq 2^{O(L)}$, it is sufficient to assume that $s_i^0 \geq 2^{-O(L)}$ and $b^T y(\mu_0) - b^T y^0 \leq n\mu_0 L$. This can easily be verified: Since $s(\mu_0)$ is dual feasible, it can be written as a convex combination of basic feasible solutions. The coordinates s_i of each basic feasible solution satisfy $s_i \leq 2^L$, and hence $s_i(\mu_0) \leq 2^L$. Therefore

$$\phi_B(y^0, \mu_0) - \phi_B(y(\mu_0), \mu_0) = \frac{b^T y(\mu_0) - b^T y^0}{\mu_0} - \sum_{i=1}^{n} \ln s_i^0 + \sum_{i=1}^{n} \ln s_i(\mu_0)$$

$$\leq O(nL).$$

Comparison with other papers

Short–step path–following were proposed and analyzed in Gonzaga [47] and Roos and Vial [133] for linear programming and in Ye [158], Ben Daya and Shetty [12], and Goldfarb and Liu [44] for quadratic programming. All these methods require $O(\sqrt{n}L)$ iterations.

It is interesting to compare the δ–measure with other measures used in the literature. Gonzaga [47] showed for his algorithm that in each iteration

$$\|S(\mu)^{-1}s - e\| \leq 0.015,$$

where s is the current iterate. Note that this distance measure is not computable since we do not know $y(\mu)$. This distance measure is related to the δ–measure as follows:

$$\|S(\mu)^{-1}s - e\| = \|\frac{Sx(\mu)}{\mu} - e\| \leq \delta(y, \mu),$$

where the equality follows from $S(\mu)x(\mu) = \mu e$ (2.9) and the inequality follows from the definition of $\delta(y, \mu)$ (2.12).

Long–step path–following methods were analyzed in Gonzaga [51] and Roos and Vial [132] for linear programming and Nesterov and Nemirovsky [119] for quadratic programming. In [132] only the $O(nL)$ iteration bound (for linear programming) is obtained for the long–step version. A disadvantage of their analysis is that, since they use global upper bounds for the slack variables (i.e. $s_i \leq 2^L$), the iteration bound is $O(nL)$ for all values of the accuracy parameter ϵ. The iteration bounds obtained for linear programming by Gonzaga in [51] are the same as obtained here, but many of our lemmas are sharper and the proofs are simpler, since we use the quadratic convergence result. E.g. in [51], although carried out in the primal formulation, similar results as Lemma 2.4 and 2.5 have been obtained in a different way, for more centered y, namely $\delta(y, \mu) \leq 0.1$, while our results hold for $\delta(y, \mu) < 1$. Moreover, for $\delta(y, \mu) \leq 0.1$ our Lemma 2.5 is tighter; for $\delta \leq 0.1$, we obtain from Lemma 2.5 that $|b^T y - b^T y(\mu)| \leq 0.13\mu\sqrt{n}$, whereas in [51] $|b^T y - b^T y(\mu)| \leq 0.20\mu\sqrt{n}$ is obtained.

In Chapter 5 of their monograph [119], Nesterov and Nemirovsky analyzed certain long–step barrier methods for linearly constrained convex quadratic programming problems. Their analysis is totally different from ours: it is not based on changes in the barrier function value, for example. From their (very complicated) analysis it can be extracted that the total number of iterations is at most $O((\frac{1}{\theta} + n^4\theta^7)L \ln n)$. Note that the iteration bound given in Theorem 2.7 is better than this one. Our bound is much better if we deal with real long–step algorithms (i.e. θ is large). For example:

- If we take long–steps then Nesterov and Nemirovsky require $O(n^4 L \ln n)$ iterations. This bound is much worse than our $O(nL)$ iteration bound;

- If we take medium–steps then Nesterov and Nemirovsky require $O(\sqrt{n}L \ln n)$ iterations. The difference with our $O(\sqrt{n}L)$ iteration bound is less significant in this case.

Kortanek and Zhu [84] extended the results for convex quadratic programming (obtained in Section 2.4) to linearly constrained convex programming, where the convex objective function satisfies the so–called Scaled Lipschitz Condition. Using the analysis of Section 2.4, Anstreicher [4] obtained the interesting result that the logarithmic barrier method, as implemented in the SUMT–3 and SUMT–4 codes, solves linear and convex quadratic programs in $O(\sqrt{n}L \ln L)$ iterations, with proper initialization and choice of parameters. Anstreicher [5] also analyzed an efficient way for obtaining finely centered points, i.e. points for which δ is very small.

There may be some convex programming problems which satisfy the Relative Lipschitz Condition (see Section 2.5.2) but do not satisfy the self–concordance condition (this only happens if one of the functions involved is not three times continuously differentiable). For these problems the analysis given in [29] is still valid. In that paper we obtained an $O((1 + M^2)n \ln \frac{n\mu_0}{\epsilon})$ iteration bound for the long–step logarithmic barrier method and an $O((1 + M^{\frac{5}{2}})\sqrt{n} \ln \frac{n\mu_0}{\epsilon})$ iteration bound for the medium–step variant, where M is the Relative Lipschitz constant. In Section 2.5 we even obtained a better complexity, namely $O(\kappa\sqrt{n} \ln \frac{n\mu_0}{\epsilon})$.

We remark that in [26] we have proved a more general result than Theorem 2.8. There we also proved that the dual objective along the paths defined by inverse barrier functions is also monotonically increasing. This was proved by using the 'dual' function of an inverse barrier function, which is a quasi barrier function, introduced by Hamala [58]. In [26] we also proved the monotonicity of the objective function along the paths defined by the quasi barrier functions.

Linear Complementarity Problems

It is well–known (see e.g. Murty [116]) that the Linear Complementarity Problem (LCP)

$$(LCP) \qquad \begin{cases} y = Rx + v, \ x,y \geq 0 \\ x^T y = 0, \end{cases}$$

where matrix R and vector v are given, is completely equivalent to the following quadratic programming problem

$$\begin{cases} \min \ x^T y \\ y = Rx + v, \ x,y \geq 0. \end{cases}$$

If R is positive semi–definite, then this problem is equivalent to a convex quadratic programming problem, since

$$\begin{aligned} x^T y &= x^T Rx + v^T x \\ &= \frac{1}{2}x^T (R + R^T)x + v^T x \\ &= \frac{1}{2}x^T Qx + v^T x, \end{aligned}$$

where $Q = R + R^T$. Consequently, the algorithm analyzed in Section 2.4 for quadratic programming, can also be applied to positive semi–definite LCP's.

Vector of barrier parameters

The logarithmic barrier function $\phi_{\mathrm{B}}(y, \mu)$ studied in the previous sections has only one barrier parameter: to each constraint the same barrier parameter value is associated. Let us now introduce a different logarithmic barrier function, by associating a different barrier parameter μ_i to each constraint:

$$\phi(y, \mu) = -f_0(y) - \sum_{i=1}^{n} \mu_i \ln(-f_i(y)),$$

where μ now denotes the barrier parameter vector.

In the sequel we will concentrate on the linear programming case. Extensions to convex quadratic and smooth convex programming can also be made.

There are several reasons for using different barrier parameters. First, as we will see below, given interior solutions for $(\mathcal{LP})$ and $(\mathcal{LD})$ we can define a barrier parameter vector such that the given dual solution minimizes $\phi(y, \mu)$. Note that in the analysis for the logarithmic barrier method we assumed that the initial point lies close to the initial reference point on the path. This can be accomplished by a transformation of the original problem (see e.g. [127, 111]), which increases the dimensions and (in the linear programming case) destroys the possible special structure of the constraint matrix A. Hence, avoiding such a transformation is useful.

Another reason for using different barrier parameters, is that this intrinsically rescales the problem in favour of the constraints with the largest barrier parameter. Thus, if the user has any prior judgements regarding the likelihood of particular constraints being active in the optimal solution, this judgement can be easily and systematically incorporated into the algorithm.

Finally, different barrier parameters can also be used to make the calculations more stable. Since the central path is an analytic concept (and not geometric), it can happen that the central path approaches its limit point along the boundary of the feasible region, which can cause numerical difficulties. Manipulating the different barrier parameters during the process of the algorithm, we can force the iterates to approach its limit point more from within.

The necessary first order optimality conditions for the unique minimum of $\phi(y, \mu)$ are:

$$\begin{cases} A^T y + s = c, & s \geq 0 \\ Ax = b, & x \geq 0 \\ Xs = \mu. \end{cases} \tag{2.90}$$

Consequently, if we have available an interior primal and dual solution pair (x^0, y^0), then defining $\mu^0 = X^0 s^0$ gives that y^0 minimizes $\phi(y, \mu^0)$.

Roos and Den Hertog [130] proposed and analyzed a short–step method based on such barrier vectors. Given the initial value μ^0 for the barrier parameter vector, they reduce each barrier parameter with the same factor $1 - \theta$, whereafter a unit Newton step is carried out to return to the new reference point. It was shown that after at most

$$O(\omega\sqrt{n}\ln\frac{n\max_i \mu_i^0}{\epsilon}) \tag{2.91}$$

iterations an ϵ–optimal solution can be found, where

$$\omega = \frac{\max_i \mu_i^0}{\min_i \mu_i^0}. \tag{2.92}$$

This analysis was extended to quadratic programming in [23].

Now we will give another method, which has a better complexity bound[13]. Without loss of generality, we assume that

$$\mu_1^0 \leq \mu_2^0 \leq \cdots \leq \mu_n^0.$$

Before reducing all the barrier parameters with the same factor, we first increase μ_1^0 repeatedly with a factor $1+\theta$, until we reach the value of μ_2^0. Then we increase both μ_1^0 and μ_2^0 (which are equal at this point) simultaneously, until we reach μ_3^0, and so on. Finally, we end up with the situation that all barrier parameters are equal to μ_n^0. This means that from this point we can start the 'normal' logarithmic barrier method, since we then have a point in the vicinity of $y(\mu_n^0)$, which is on the central path.

Modifying the analysis in Section 2.3, it can be proved that the short– and medium–step variant need $O(\sqrt{n}\ln\omega)$ Newton iterations, and the long–step variant $O(n\ln\omega)$ iterations to obtain a point y such that $\delta(y,\mu_n^0) \leq \frac{1}{2}$ (ω is given by (2.92)). From this point the normal logarithmic barrier method needs $O(\sqrt{n}\ln\frac{n\mu_n^0}{\epsilon})$ and $O(n\ln\frac{n\mu_n^0}{\epsilon})$ Newton iterations, respectively, to obtain an ϵ–optimal solution. Consequently, these complexities allow us to start from an initial feasible point for which the coordinates are between $\Theta(\epsilon)$ and $\Theta(\frac{1}{\epsilon})$, contrary to the complexity bound (2.91) obtained in [130]. E.g., for $\epsilon = 2^{-2L}$ this means that the algorithm can start 'almost everywhere'.

---

[13]We learned from Osman Güler that a similar idea is worked out by Mizuno [104] for the primal–dual path–following method.

Chapter 3

The center method

In this chapter we analyze the complexity of special variants of the center method for some classes of convex problems. We will deal with so-called long–, medium– and short–step methods. First, we will give a general framework for the center method. Then we will give the complexity analysis for linear programming and a class of convex programming problems, respectively. The self–concordance property introduced by Nesterov and Nemirovsky [119] plays again a key role in the analysis.

For the complexity analysis of the center method, we can use many of the results obtained for the logarithmic barrier method in Chapter 2, due to the close relationship between the logarithmic barrier method and the center method. For example, the quadratic convergence property for linear programming is an immediate consequence of Lemma 2.3. Of course there are important differences, especially in the analysis for obtaining upper bounds for the number of inner and outer iterations. The complexity bounds for the center method, obtained in this chapter, are comparable with those of the logarithmic barrier method.

We note that we do not treat the linearly constrained convex quadratic programming case in this chapter, since the above mentioned relationship does not hold in this case (the center method for this problem is related to the logarithmic barrier method for *quadratically* constrained convex quadratic programming, for which we do not have a separate analysis in Chapter 2).

3.1 General framework

Recall the convex programming problem $(\mathcal{CP})$ as given in Section 2.1:

$$(\mathcal{CP}) \quad \begin{cases} \max\ f_0(y) \\ f_i(y) \leq 0, \quad i = 1, \cdots, n \\ y \in \mathbb{R}^m, \end{cases}$$

and its Wolfe [152] dual problem

$$(\mathcal{CD}) \quad \begin{cases} \min\ f_0(y) - \sum_{i=1}^{n} x_i f_i(y) \\ \sum_{i=1}^{n} x_i \nabla f_i(y) = \nabla f_0(y) \\ x_i \geq 0. \end{cases}$$

We make the same assumptions as in Section 2.1 (Assumptions 1–3). We associate the following *distance function* with $(\mathcal{CP})$

$$\phi_{\mathrm{D}}(y, z) = -q \ln(f_0(y) - z) - \sum_{i=1}^{n} \ln(-f_i(y)), \tag{3.1}$$

where z is a lower bound for the optimal value z^*, and q a given positive integer, which will be discussed later on. This distance function was introduced by Huard [60].

For the classes of problems we will consider in the next sections (linear programming and convex programming satisfying the self–concordance condition) we have that $\phi_{\mathrm{D}}(y, z)$ is strictly convex on its domain $\mathcal{F}^0$. So, in the remainder of this section we may assume that $\phi_{\mathrm{D}}(y, z)$ is strictly convex. It also takes infinite values on the boundary of the feasible set. Hence this distance function achieves the minimal value in its domain (for fixed z) at a unique point, which is denoted by $y(z)$. The necessary and sufficient Karush–Kuhn–Tucker conditions for this minimum are:

$$\begin{cases} f_i(y) \leq 0, & 1 \leq i \leq n, \\ \sum_{i=1}^{n} x_i \nabla f_i(y) = \nabla f_0(y), & x \geq 0, \\ -f_i(y)x_i = \frac{f_0(y)-z}{q}, & 1 \leq i \leq n. \end{cases} \tag{3.2}$$

Comparing this system of equations with (2.5) it is easy to verify that $y(z)$ lies on the so–called central path of the problem, which was introduced in Section 2.1. To be more precise: from (2.5) and (3.2) it follows that $y(z) = y(\mu)$ for $\mu = \frac{f_0(y(z))-z}{q}$. Consequently, the logarithmic barrier function (2.1) and the distance function (3.1) yield two different parameterizations of the same central path. In Section 3.2 we will give some explicit examples.

We can rewrite $\phi_{\mathrm{D}}(y, z)$ as

$$\phi_{\mathrm{D}}(y, z) = -\sum_{i=1}^{n+q} \ln(-f_i(y)), \tag{3.3}$$

where $-f_i(y) = f_0(y) - z$ for $n + 1 \leq i \leq n + q$. Moreover, we introduce $\mathcal{F}_z$ which is the bounded convex region defined by the constraint functions and the objective constraint $f_0(y) - z \geq 0$, which is replicated q times. More precisely

$$\mathcal{F}_z = \{y\ :\ f_i(y) \leq 0,\ 1 \leq i \leq n + q\}.$$

Let $\mathcal{F}_z^0$ denote the interior of $\mathcal{F}_z$. Now we define the concept of the so–called analytic center of the bounded convex region $\mathcal{F}_z$.

Definition: The *analytic center* of the bounded convex region $\mathcal{F}_z$ is the point which maximizes

$$\prod_{i=1}^{n+q}(-f_i(y)), \tag{3.4}$$

the product of the slack values.

This concept was introduced and studied by Sonnevend [135]. Since $y(z)$ maximizes

$$e^{-\phi_D(y,z)} = \prod_{i=1}^{n+q}(-f_i(y)),$$

it easily follows that $y(z)$ is the analytic center of $\mathcal{F}_z$.

Using this observation, it immediately follows that (3.2) is equivalent to the necessary and sufficient Karush–Kuhn–Tucker conditions for (3.4):

$$\begin{cases} f_i(y) \leq 0, & 1 \leq i \leq n+q \\ \sum_{i=1}^{n+q} \tilde{x}_i \nabla f_i(y) = 0, & \tilde{x} \geq 0 \\ -f_i(y)\tilde{x}_i = 1, & 1 \leq i \leq n+q. \end{cases} \tag{3.5}$$

Note that $\phi_D(y, z)$ is in fact the logarithmic barrier function for the following problem

$$\max \quad \{0 \ : \ y \in \mathcal{F}_z\}.$$

Hence, for the analysis of the method of centers, we can use many results obtained for the logarithmic barrier function method in Chapter 2.

It is important to know that the analytic center is an analytic concept and not a geometric one. As a consequence, the analytic center depends also on the description of the feasible region. The effect of replicating a constraint on the position of the analytic center is shown in Figure 3.1. This explains why the objective constraint is replicated q times, since then the resulting analytic center is pushed in the direction of the optimum. The complexity bounds will appear to depend on this q, and we will derive the best value from the complexity point of view.

Loosely speaking, the method of centers works as follows. Given a lower bound z for the optimal value, we try to reach the vicinity of $y(z)$, the analytic center of the current feasible region. Then we increase the lower bound, which means that the q objective constraints are shifted, and we try to reach the vicinity of the new center.

In the method of centers we need the first and second order derivatives of $\phi_D(y, z)$, which are given by

$$g(y, z) := \nabla \phi_D(y, z) = \sum_{i=1}^{n+q} \frac{\nabla f_i(y)}{-f_i(y)}$$

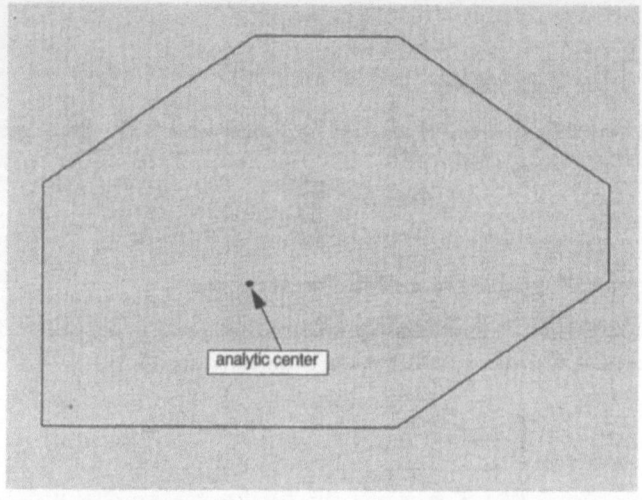

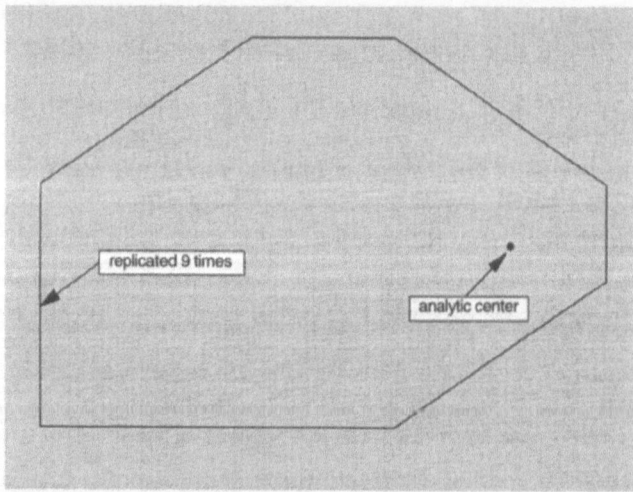

Figure 3.1: The effect of replicating a constraint on the position of the analytic center of a polytope.

and

$$H(y, z) := \nabla^2 \phi_{\mathrm{D}}(y, z)$$

$$= \sum_{i=1}^{n+q} \left[\frac{\nabla^2 f_i(y)}{-f_i(y)} + \frac{\nabla f_i(y) \nabla f_i(y)^T}{f_i(y)^2} \right]. \tag{3.6}$$

If no confusion is possible we will write g and H instead of $g(y, z)$ and $H(y, z)$, for briefness' sake. We will use the measure $\|.\|_H$ to measure distances between points, and especially the distance to $y(z)$. The definition of this measure is as follows:

$$\|v\|_H = \sqrt{v^T H v}.$$

Because H is positive definite, $\|.\|_H$ defines a norm.

As for the logarithmic barrier function, there are some important issues. To find (an iterate close to) the analytic center, which is in fact equivalent to minimizing $\phi_{\mathrm{D}}(y, z)$, we use again Newton's method with (approximate) line search procedures. Note that the Newton direction is given by

$$p(y, z) = -H(y, z)^{-1} g(y, z) = -H^{-1} g.$$

If no confusion is possible we will write, for shortness sake, p instead of $p(y, z)$. The proximity criterion for this minimizing process is again the Hessian norm of the search direction. More precisely: we stop the minimizing procedure if $\|p\|_H \le \tau$, where τ is a certain tolerance. (Note that $\|p\|_H = 0$ if and only if $y = y(z)$). If the proximity criterion is satisfied, we update the lower bound z as follows: $\bar{z} := z + \theta(f_0(y) - z)$, for some $0 < \theta < 1$. Note that $\bar{z}$ is really a lower bound, because $\bar{z} < f_0(y) \le z^*$.

We now describe the algorithm to find an ϵ-optimal solution.

Center Algorithm

Input:
ϵ is the accuracy parameter;
τ is the proximity tolerance;
θ is the updating factor, $0 < \theta < 1$;
$z^0 < f_0(y^0)$ is a lower bound for the optimal value;
y^0 is a given interior feasible point such that $\|p(y^0, z^0)\|_{H(y^0, z^0)} \le \tau$;

begin
    $y := y^0$; $z := z^0$; $\Delta = 4(1 + \frac{n}{q})$;
    while $f_0(y) - z > \frac{\epsilon}{\Delta}$ **do**
    begin (outer step)
        $z := z + \theta(f_0(y) - z)$;
        while $\|p\|_H > \tau$ **do**

      begin (inner step)
        $\tilde{\alpha} := \arg\min_{\alpha>0} \{\phi_{\mathrm{D}}(y + \alpha p, z) : y + \alpha p \in \mathcal{F}^0\}$
        $y := y + \tilde{\alpha} p$
      end (end inner step)
    end (end outer step)
end.

In the analysis we will show that taking an appropriate steplength is sufficient for the complexity analysis. For finding an initial point that satisfies the input assumptions of the algorithm, we refer the reader to Renegar [127], Monteiro and Adler [110] and Güler et al. [57] for linear programming, Ye [158] for quadratic programming, and Jarre [66] and Mehrotra and Sun [99] for convex programming. Later on the 'centering assumption' will be alleviated.

Analogously to Section 2.1, we introduce some terminology. We call the algorithm a

- **long–step** algorithm if θ is a constant $(0 < \theta < 1)$, independent of n and ϵ;

- **medium–step** algorithm if $\theta = \frac{\nu}{\sqrt{n}}$, where $\nu > 0$ is an arbitrary constant, possibly large, and independent of n and ϵ;

- **short–step** algorithm if $\theta = \frac{\nu}{\sqrt{n}}$, and ν is so small (e.g. $\frac{1}{8}$), that after the update of z one unit Newton step is sufficient to reach the vicinity of the new center.

In this chapter we will analyze such long–, medium– and short–step algorithms.

3.2   Centers for some examples

In this section we will illustrate the theory of centers by calculating centers for some explicit examples. To compare the results we will use the same examples as in Section 2.2. In all the examples we will take $q = n$.

Example 1. Consider the following simple linear programming problem ($m = 2$ and $n = 4$):

$$\begin{cases} \max 2y_1 + y_2 \\ y_1 \leq 1 \\ -y_1 \leq 1 \\ y_2 \leq 1 \\ -y_2 \leq 1. \end{cases}$$

The distance function for this problem is:

$$\phi_D(y, z) = -4\ln(2y_1 + y_2 - z) - \ln(1 - y_1)$$
$$- \ln(1 - y_2) - \ln(1 + y_1) - \ln(1 + y_2).$$

This function is minimized by setting the derivatives (with respect to y_1 and y_2) equal to zero:

$$-\frac{8}{2y_1 + y_2 - z} + \frac{1}{1 - y_1} - \frac{1}{1 + y_1} = 0$$

and

$$-\frac{4}{2y_1 + y_2 - z} + \frac{1}{1 - y_2} - \frac{1}{1 + y_2} = 0.$$

It is difficult to get explicit expressions for $y_1(z)$ and $y_2(z)$. However, multiplying the second equality by 2 and subtracting the two equations, and rewriting the result we can eliminate z and get the following result:

$$y_2(z) = y_1(z) - \frac{1}{y_1(z)} + \sqrt{y_1(z)^2 - 1 + \frac{1}{y_1(z)^2}}. \qquad (3.7)$$

It can easily be verified that $(0,0)$ and $(1,1)$ are on the path of centers. Moreover, the μ–centers for this example,

$$y_1(\mu) = \frac{1}{2}\sqrt{\mu^2 + 4} - \frac{1}{2}\mu$$

and

$$y_2(\mu) = \sqrt{\mu^2 + 1} - \mu,$$

calculated in Example 1 of Section 2.2 also satisfy (3.7). This means that the two paths are identical.

Example 2. Consider again Example 1 but now with objective function y_2 instead of $2y_1 + y_2$. It is clear that the optimal solution is not unique for this example. Doing similar algebraic manipulations as in Example 1, we can now get explicit expressions for $y_1(z)$ and $y_2(z)$:

$$y_1(z) = 0$$

and

$$y_2(z) = \frac{1}{6}z + \frac{1}{6}\sqrt{z^2 + 24}.$$

Note that this path of centers coincides with the central path of Example 2 in Section 2.2. The optimal point can be determined by using the property that in an optimal solution we have $z = f_0(y(z))$ or, equivalently in this case

$$z = \frac{1}{6}z + \frac{1}{6}\sqrt{z^2 + 24}.$$

Solving for z we get $z = 1$. Hence the optimal value is 1 and an optimal solution is $(y_1(1), y_2(1)) = (0, 1)$. The path of centers ends in $(0, 1)$, the analytic center of the optimal facet.

Example 3a. Let us now consider the problem

$$
\begin{cases}
\max 2y_1 + y_2 \\
y_1 \le 1 \\
y_2 \le 1
\end{cases}
$$

By doing the same calculations as for Examples 1 and 2 we obtain that

$$
y_1(z) = \frac{5}{8} + \frac{1}{8}z
$$

and

$$
y_2(z) = \frac{1}{4} + \frac{1}{4}z.
$$

The optimal value of this problem can be calculated from $z = f_0(y(z))$, which yields $z = 3$. Consequently, the optimal point of this problem is $(y_1(3), y_2(3)) = (1,1)$. Hence the path of centers is a line ending in the optimal solution $(1,1)$. Of course, the same path is obtained in Example 3a of Section 2.2.

Example 3b. As shown in Example 3b. of Section 2.2, the observation made in Example 3a (the central path is a line) is true also in a more general setting: if the feasible region of a linear program is a cone then the central path is a line, ending in the top of the cone. We will show that this is also true for the path of centers. Again, suppose that the constraints of the linear problem describe a cone in $\mathbb{R}^m$. Moreover, let the objective function be such that the top of the cone is the optimal solution. Let the constraint matrix be denoted by A^T, which is a square and invertible $m \times m$ matrix, and let b, an m-dimensional vector be the objective vector. So the problem is as follows

$$
\begin{cases}
\max b^T y \\
A^T y \le c.
\end{cases}
$$

Again, note that $t = A^{-T}c$ is the top of the cone. The distance function for this problem is

$$
\phi_D(y, z) = -q \ln(b^T y - z) - \sum_{i=1}^{m} \ln(c_i - a_i^T y),
$$

where a_i is the i-th row of A^T. It can easily be verified that (3.2) reduces to[1]

$$
\begin{cases}
A^T y + s = c, \quad s \ge 0 \\
Ax = b, \qquad\quad x \ge 0 \\
Sx = \frac{b^T y - z}{q} e.
\end{cases}
\tag{3.8}
$$

---

[1]See also Section 3.3.

Now we can calculate $y(z)$ in a similar way as we calculated $y(\mu)$ in Example 3b. of Section 2.2:

$$y(z) = A^{-T}(c - \frac{b^T y - z}{q} v),$$

where

$$v_i = \frac{1}{(A^{-1}b)_i}.$$

Equivalently we have

$$\left(I + A^{-T}\frac{vb^T}{q}\right)y = A^{-T}c + \frac{z}{q}A^{-T}v.$$

Consequently, the path of centers is a line. It can easily be verified that this line ends in the optimal point, namely the top of the cone $(t = A^{-T}c)$. Moreover, one can verify that this line is the same as the line calculated in Example 3b. of Section 2.2.

Example 4. Let us now consider the following simple convex quadratic programming problem ($m = 2$ and $n = 1$):

$$\begin{cases} \max -y_1 - y_2 \\ y_1^2 + 2y_2^2 \leq 1 \end{cases}$$

The distance function for this problem is:

$$\phi_D(y, z) = -\ln(-y_1 - y_2 - z) - \ln(1 - y_1^2 - 2y_2^2).$$

This function is minimized by setting the derivatives (with respect to y_1 and y_2) equal to zero. Solving these equations we obtain

$$y_1(z) = -\frac{2}{9}z - \frac{1}{9}\sqrt{4z^2 + 18}$$

and

$$y_2(z) = -\frac{1}{9}z - \frac{1}{18}\sqrt{4z^2 + 18}.$$

To calculate the endpoint of this path of centers we first calculate the optimal value of the problem by setting

$$z = f_0(y(z)),$$

from which we obtain

$$z = \sqrt{\frac{3}{2}}.$$

Substituting this value into the formulas for $y_1(z)$ and $y_2(z)$ we obtain that the path of centers ends in $(-\frac{1}{3}\sqrt{6}, -\frac{1}{6}\sqrt{6})$. Moreover, this path of centers starts in $(0,0)$, the analytic center of the feasible region, since

$$\lim_{z \to -\infty} y_1(z) = \lim_{z \to -\infty} y_2(z) = 0.$$

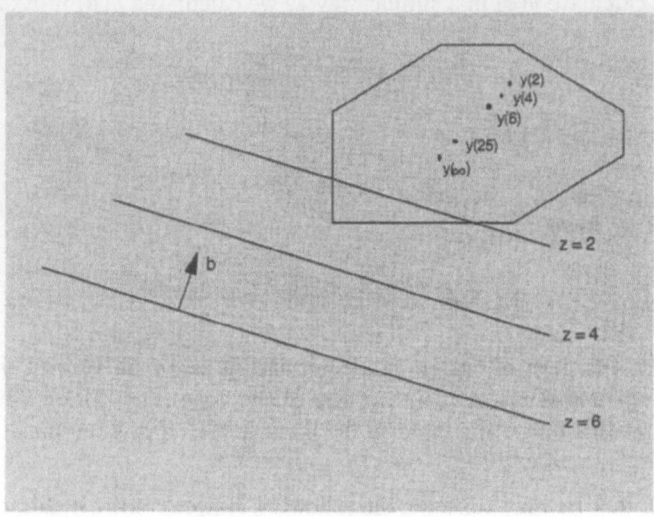

Figure 3.2: Several centers for different values of z.

The same results were obtained in Section 2.2 for the central path of the same problem.

Example 5. For the above three examples we were able to give an explicit expression for the central path. This is, of course, in general not possible. The following linear programming problem, with $m = 2$ and $n = 7$ is an example of this:

$$
\begin{cases}
\max \ y_1 + 2y_2 \\
y_1 \leq 1 \\
-y_1 \leq 1 \\
y_2 \leq 1 \\
-y_2 \leq 1 \\
-y_1 + y_2 \leq 1.25 \\
y_1 + y_2 \leq 1.25 \\
y_1 - y_2 \leq 1.25
\end{cases}
$$

It can easily be verified for this example that it is not possible to give an explicit expression for $y_1(z)$ and $y_2(z)$. In Figure 3.2 some centers for different values of the lower bound are shown. Note that they are on the central path. In this figure the 'level constraints' are replicated 7 times ($q = n = 7$).

3.3   Linear programming

3.3.1   Properties at and near the centers

Consider the dual linear programming problem ($\mathcal{LD}$) making the same assumptions as in Section 2.3. The distance function for this problem is given by

$$\phi_{\mathrm{D}}(y, z) = -q \ln(b^T y - z) - \sum_{i=1}^{n} \ln s_i,$$

where q is a (positive) integer.

The Hessian of this function,

$$\frac{q}{(b^T y - z)^2} b b^T + A S^{-2} A^T,$$

is positive definite since we assumed that A has full rank. Hence, the distance function $\phi_{\mathrm{D}}(y, z)$ achieves a minimum value at a unique point, denoted as $y(z)$. The necessary and sufficient Karush–Kuhn–Tucker conditions for $y(z)$ are:

$$\begin{cases} A^T y + s = c, & s \geq 0 \\ A x = b, & x \geq 0 \\ X s = \frac{b^T y - z}{q} e, \end{cases} \qquad (3.9)$$

where x is an n–dimensional vector. The unique solution of this system is denoted by $(x(z), y(z), s(z))$. As already mentioned in Section 3.1, $y(z)$ lies on the central trajectory of problem ($\mathcal{LD}$). More precisely, comparing (3.9) and (2.9) we have $y(\mu) = y(z)$ for

$$\mu = \frac{b^T y - z}{q}.$$

As explained in Section 3.1, the minimizing point $y(z)$ can be considered as the analytic center of $\mathcal{F}_z$, which is defined as

$$\{y \ : \ A^T y \leq c, \underbrace{-b^T y \leq -z, \cdots, -b^T y \leq -z}_{q \text{ times}} \}.$$

Note that $\mathcal{F}_z$ is bounded, since we assumed that the optimal set is bounded. Now, let matrix $\tilde{A}$ be given by

$$\tilde{A} := (a_1, \cdots, a_n, a_{n+1}, \cdots, a_{n+q}),$$

where $a_i := -b$ for $n+1 \leq i \leq n+q$. The components of the diagonal matrix $\tilde{S}$ are defined as s_i for $1 \leq i \leq n$ and $b^T y - z$ for $n+1 \leq i \leq n+q$. The components of the vector $\tilde{c}$ are defined as c_i for $1 \leq i \leq n$ and $-z$ for $n+1 \leq i \leq n+q$. Consequently,

$$\phi_{\mathrm{D}}(y, z) = - \sum_{i=1}^{n+q} \ln \tilde{s}_i. \qquad (3.10)$$

Using this notation, we observe that $y(z)$ is the analytic center of

$$\mathcal{F}_z = \{y \ : \ \tilde{A}^T y \le \tilde{c}\},$$

and (3.9) is equivalent to the Karush–Kuhn–Tucker conditions for (3.10):

$$\begin{cases} \tilde{A}^T y + \tilde{s} = \tilde{c}, & \tilde{s} \ge 0 \\ \tilde{A}\tilde{x} = 0, & \tilde{x} \ge 0 \\ \tilde{X}\tilde{s} = e. \end{cases} \tag{3.11}$$

The unique solution of this system is denoted by $(\tilde{x}(z), y(z), \tilde{s}(z))$.

The following lemma states that the dual objective $b^T y(z)$ increases, the primal objective $c^T x(z)$ and $b^T y(z) - z$ decrease, if z increases.

Lemma 3.1 *The dual objective $b^T y(z)$ is monotonically increasing, the primal objective $c^T x(z)$ and $b^T y(z) - z$ are monotonically decreasing if z increases.*

Proof: We first prove the last part of the lemma. Suppose $\bar{z} > z$, such that $b^T y(z) > \bar{z}$. Consider the function

$$\phi_{\mathrm{D}}^d(\tilde{x}, z) = \tilde{c}^T \tilde{x} - \sum_{i=1}^{n+q} \ln \tilde{x}_i - (n+q).$$

It is easy to verify that $\tilde{x}(z)$ is the minimum of $\phi_{\mathrm{D}}^d(\tilde{x}, z)$ over the region $\{\tilde{x} \ : \ \tilde{A}\tilde{x} = 0, \ \tilde{x} \ge 0\}$. (The Karush–Kuhn–Tucker conditions for $\phi_{\mathrm{D}}^d(\tilde{x}, z)$ are exactly (3.11).) Note that, due to (3.11), the last q coordinates of $\tilde{x}(z)$ are equal to $\tilde{x}_{n+1}(z)$. The first n components of $\tilde{x}(z)$ are denoted as the vector $\hat{x}(z)$. Now, since $\tilde{x}(z)$ minimizes $\phi_{\mathrm{D}}^d(\tilde{x}, z)$ and $\tilde{x}(\bar{z})$ minimizes $\phi_{\mathrm{D}}^d(\tilde{x}, \bar{z})$, we have

$$\phi_{\mathrm{D}}^d(\tilde{x}(z), z) \le \phi_{\mathrm{D}}^d(\tilde{x}(\bar{z}), z)$$

and

$$\phi_{\mathrm{D}}^d(\tilde{x}(\bar{z}), \bar{z}) \le \phi_{\mathrm{D}}^d(\tilde{x}(z), \bar{z})$$

or equivalently,

$$\begin{aligned} c^T \hat{x}(z) \quad &- \quad qz\tilde{x}_{n+1}(z) - \sum_{i=1}^{n+q} \ln \tilde{x}_i(z) - (n+q) \\ &\le \quad c^T \hat{x}(\bar{z}) - qz\tilde{x}_{n+1}(\bar{z}) - \sum_{i=1}^{n+q} \ln \tilde{x}_i(\bar{z}) - (n+q), \end{aligned}$$

and

$$\begin{aligned} c^T \hat{x}(\bar{z}) \quad &- \quad q\bar{z}\tilde{x}_{n+1}(\bar{z}) - \sum_{i=1}^{n+q} \ln \tilde{x}_i(\bar{z}) - (n+q) \\ &\le \quad c^T \hat{x}(z) - q\bar{z}\tilde{x}_{n+1}(z) - \sum_{i=1}^{n+q} \ln \tilde{x}_i(z) - (n+q). \end{aligned}$$

Adding the two inequalities gives

$$-q\left(z\tilde{x}_{n+1}(z) + \bar{z}\tilde{x}_{n+1}(\bar{z})\right) \le -q\left(z\tilde{x}_{n+1}(\bar{z}) + \bar{z}\tilde{x}_{n+1}(z)\right),$$

or equivalently,

$$(z - \bar{z})\left(\tilde{x}_{n+1}(z) - \tilde{x}_{n+1}(\bar{z})\right) \ge 0.$$

Since $\bar{z} > z$, this means that

$$\tilde{x}_{n+1}(z) \le \tilde{x}_{n+1}(\bar{z}). \tag{3.12}$$

Now from (3.11) we have

$$\tilde{x}_{n+1}(z) = \frac{1}{\tilde{s}_{n+1}(z)} = \frac{1}{b^T y(z) - z},$$

and a similar expression for $\tilde{x}_{n+1}(\bar{z})$. Substituting this into (3.12) gives that $b^T y(z) - z$ monotonically decreases. To prove that $c^T x(z)$ is monotonically decreasing and $b^T y(z)$ is monotonically increasing, observe that $x(z) = x(\mu)$, where $\mu = \frac{b^T y(z) - z}{q}$. Since $b^T y(z) - z$ is monotonically decreasing, it follows from Lemma 2.1 that $c^T x(\mu)$ and thus $c^T x(z)$ is monotonically decreasing and that $b^T y(\mu)$ and thus $b^T y(z)$ is monotonically increasing.

There is also a direct proof for the monotonicity of $b^T y(z)$. Since $y(z)$ minimizes $\phi_\mathrm{D}(y, z)$ and $y(\bar{z})$ minimizes $\phi_\mathrm{D}(y, \bar{z})$ we obviously have

$$\phi_\mathrm{D}(y(z), z) \le \phi_\mathrm{D}(y(\bar{z}), z)$$

and

$$\phi_\mathrm{D}(y(\bar{z}), \bar{z}) \le \phi_\mathrm{D}(y(z), \bar{z}),$$

or equivalently,

$$-q\ln(b^T y(z) - z) - \sum_{i=1}^{n} \ln s_i(z) \le -q\ln(b^T y(\bar{z}) - z) - \sum_{i=1}^{n} \ln s_i(\bar{z}),$$

and

$$-q\ln(b^T y(\bar{z}) - \bar{z}) - \sum_{i=1}^{n} \ln s_i(\bar{z}) \le -q\ln(b^T y(z) - \bar{z}) - \sum_{i=1}^{n} \ln s_i(z).$$

Adding the two inequalities gives

$$-q\ln(b^T y(z) - z) - q\ln(b^T y(\bar{z}) - \bar{z}) \le -q\ln(b^T y(\bar{z}) - z) - q\ln(b^T y(z) - \bar{z}),$$

or equivalently, by taking the exponential for both sides

$$-(b^T y(z) - z)(b^T y(\bar{z}) - \bar{z}) \le -(b^T y(\bar{z}) - z)(b^T y(z) - \bar{z}).$$

This reduces to

$$zb^T y(\bar{z}) + \bar{z}b^T y(z) \le zb^T y(z) + \bar{z}b^T y(\bar{z}),$$

or

$$(z - \bar{z})(b^T y(\bar{z}) - b^T y(z)) \le 0.$$

Since $\bar{z} > z$ this means that $b^T y(z) \le b^T y(\bar{z})$. □

The following lemma gives an upper bound for the gap $z^* - z$.

Lemma 3.2 *One has*

$$z^* - z \le (1 + \frac{n}{q})(b^T y(z) - z).$$

Proof: The exact center $y(z)$ minimizes the distance function for z. The necessary and sufficient conditions for $y(z)$ are (3.9). From these conditions we derive that $Ax(z) = b$ and $x \ge 0$. Consequently, $x(z)$ is primal feasible. Moreover, using $z^* \le c^T x(z)$ it follows that

$$
\begin{aligned}
z^* - b^T y(z) &\le c^T x(z) - b^T y(z) \\
&= x(z)^T s(z) \\
&= \frac{n}{q}(b^T y(z) - z).
\end{aligned}
$$

Consequently,

$$(z^* - z) - (b^T y(z) - z) \le \frac{n}{q}(b^T y(z) - z).$$

This implies

$$z^* - z \le (1 + \frac{n}{q})(b^T y(z) - z).$$

$\square$

For the distance function $\phi_D(y, z)$ we can easily compute the gradient and Hessian matrix:

$$
\begin{aligned}
g(y, z) &:= \nabla \phi_D(y, z) \\
&= \frac{-q}{b^T y - z} b + A S^{-1} e \\
&= \tilde{A} \tilde{S}^{-1} e,
\end{aligned}
$$

and

$$
\begin{aligned}
H(y, z) &:= \nabla^2 \phi_D(y, z) \\
&= \frac{q}{(b^T y - z)^2} b b^T + A S^{-2} A^T \\
&= \tilde{A} \tilde{S}^{-2} \tilde{A}^T,
\end{aligned}
$$

where $\tilde{A}$ and $\tilde{s}$ are as defined before. If no confusion is possible we will write, for shortness sake, g and H instead of $g(y, z)$ and $H(y, z)$.

As shown above, finding the analytic center of $\mathcal{F}_z$ is equivalent to minimizing $\phi_D(y, z)$ (see (3.10)), which is exactly the logarithmic barrier function for the following problem

$$\max\{0 \; : \; \tilde{A}^T y + \tilde{s} = c, \; \tilde{s} \ge 0\}.$$

Now we are back in the formulation of Section 2.3. So from (2.12) we can get the corresponding measure for the distance of an interior feasible point to the center $y(z)$ of $\mathcal{F}_z$:

$$\sigma(y, z) = \min_{\tilde{x}} \left\{ \|\tilde{S}\tilde{x} - e\| \; : \; \tilde{A}\tilde{x} = 0 \right\}. \tag{3.13}$$

The unique solution of the minimization problem in the definition of $\sigma(y, z)$ is denoted by $\tilde{x}(y, z)$. So we can also write

$$\sigma(y, z) = \|\tilde{S}\tilde{x}(y, z) - e\|. \qquad (3.14)$$

Since the δ–distance of a point y in $\mathcal{F}_z$ to the center $y(z)$ is equal to $\sigma(y, z)$, many results obtained for the δ–measure can be generalized directly to σ. The Newton direction $p(y, z)$ is given by

$$
\begin{aligned}
p(y, z) &= -H^{-1}g \\
&= -(\tilde{A}\tilde{S}^{-2}\tilde{A}^T)^{-1}\tilde{A}\tilde{S}^{-1}e.
\end{aligned}
$$

Most of the lemmas we will give in this section are direct consequences of lemmas proved in Section 2.3.

Lemma 3.3 *For given y and z we have*

$$\sigma(y, z) = \|p(y, z)\|_H = \|\tilde{S}^{-1}\tilde{A}^T p(y, z)\| \geq \|S^{-1}A^T p(y, z)\|.$$

Proof: The equalities are immediate consequences of Lemma 2.2. The inequality follows easily from the definition of $\tilde{A}\tilde{S}^{-1}$:

$$
\begin{aligned}
\|\tilde{S}^{-1}\tilde{A}^T p(y, z)\|^2 &= \|S^{-1}A^T p\|^2 + q(b^T p)^2 \\
&\geq \|S^{-1}A^T p\|^2.
\end{aligned}
$$

$\square$

Lemma 3.4 *If $\sigma(y, z) < 1$ then $y^+ = y + p$ is a strictly feasible point for $\mathcal{F}_z$. Moreover,*

$$\sigma(y^+, z) \leq \sigma(y, z)^2.$$

Proof: This is an immediate consequence of Lemma 2.3. $\square$

Lemma 3.5 *If $\sigma := \sigma(y, z) < 1$ then*

$$\phi_D(y, z) - \phi_D(y(z), z) \leq \frac{\sigma^2}{1 - \sigma^2}.$$

Proof: This is an immediate consequence of Lemma 2.4. $\square$

Lemma 3.6 *If $\sigma := \sigma(y, z) < 1$ and $q \geq 2\sqrt{n}$, then*

$$b^T y(z) - z \leq \left(1 + \frac{2\sqrt{n}}{q}\frac{\sigma}{1 - \sigma}\right)(b^T y - z).$$

Proof: The proof is of the same structure as the proof of Lemma 2.5. We have $p^T g = -\sigma^2$. On the other hand

$$
\begin{aligned}
-p^T g &= p^T \left(\frac{qb}{b^T y - z} - AS^{-1}e \right) \\
&= \frac{q}{b^T y - z} b^T p - e^T S^{-1} A^T p.
\end{aligned}
$$

So we have

$$
\frac{q}{b^T y - z} b^T p = \sigma^2 + e^T S^{-1} A^T p. \tag{3.15}
$$

Using the Cauchy–Schwartz inequality, we obtain

$$
|e^T S^{-1} A^T p| \le \|S^{-1} A^T p\| \, \|e\| \le \sigma \sqrt{n}, \tag{3.16}
$$

where the last inequality follows from Lemma 3.3. From (3.15) and (3.16) we derive that

$$
\begin{aligned}
|b^T p| &\le \frac{b^T y - z}{q} (\sigma^2 + \sigma \sqrt{n}) \\
&\le \sigma(1 + \sigma) \frac{b^T y - z}{q} \sqrt{n} \\
&\le 2\sigma \frac{b^T y - z}{q} \sqrt{n}. \tag{3.17}
\end{aligned}
$$

Again, let $y^0 := y$ and let $y^1, y^2, \cdots$ denote the sequence of points obtained by repeating Newton steps, starting at y^0. According to Lemma 3.4 we have

$$
\sigma(y^k, z) \le \sigma(y^{k-1}, z)^2 \le \cdots \le \sigma(y^0, z)^{2^k} = \sigma(y, z)^{2^k}.
$$

This means that $\sigma(y^k, z) \to 0$ if $k \to \infty$. Since all y^k lie in a bounded region and $\sigma(y, z) = 0$ if and only if $y = y(z)$, we have that all limit points of the sequence are $y(z)$. Hence, the sequence of Newton iterates converges to $y(z)$. Therefore, we may write

$$
\begin{aligned}
b^T y^{k+1} - z &= b^T y^k - z + b^T p(y^k, z) \\
&\le \left(1 + 2\sigma(y^k, z) \frac{\sqrt{n}}{q} \right) (b^T y^k - z) \\
&\le \left(1 + 2\sigma^{2^k} \frac{\sqrt{n}}{q} \right) (b^T y^k - z) \\
&\le (b^T y^0 - z) \prod_{i=0}^{k} \left(1 + 2\sigma^{2^i} \frac{\sqrt{n}}{q} \right),
\end{aligned}
$$

where the first inequality follows from (3.17). In particular we have

$$
\begin{aligned}
b^T y(z) - z &\le (b^T y - z) \prod_{i=0}^{\infty} \left(1 + 2\sigma^{2^i} \frac{\sqrt{n}}{q} \right) \\
&\le \left(1 + \frac{2\sqrt{n}}{q} \frac{\sigma}{1 - \sigma} \right) (b^T y - z),
\end{aligned}
$$

where the last inequality follows from Lemma B.4. This proves the lemma.   □

3.3.2 Complexity analysis

In this section we will derive an upper bound for the total number of Newton iterations. We derive an upper bound both for the number of outer iterations (updates of the lower bound) and for the number of inner iterations for an arbitrary outer iteration. The product of these two upper bounds is an upper bound for the total number of Newton iterations. First, we will deal with the long– and medium–step variant, and finally with the short–step variant. We will set $\tau = \frac{1}{2}$ in the Center Algorithm.

Theorem 3.1 *After at most*

$$\frac{2}{\theta}(1 + \frac{n}{q}) \ln \frac{4(1 + \frac{n}{q})(z^* - z^0)}{\epsilon}$$

outer iterations, the Center Algorithm ends up with an ϵ-optimal solution of the problem.

Proof: Let z^k be the lower bound in the k-th outer iteration, and y^k the iterate at the end of k outer iterations. Using Lemmas 3.2 and 3.6 and the fact that $q \geq 2\sqrt{n}$ successively, we obtain

$$\begin{aligned}
z^* - z^{k-1} &\leq (1 + \frac{n}{q})(b^T y(z^{k-1}) - z^{k-1}) \\
&\leq (1 + \frac{n}{q})\left(1 + \frac{2\sqrt{n}}{q}\right)(b^T y^{k-1} - z^{k-1}) \\
&\leq 2(1 + \frac{n}{q})(b^T y^{k-1} - z^{k-1}).
\end{aligned} \tag{3.18}$$

Moreover,

$$\begin{aligned}
\frac{z^* - z^k}{z^* - z^{k-1}} &= \frac{z^* - (z^{k-1} + \theta(b^T y^{k-1} - z^{k-1}))}{z^* - z^{k-1}} \\
&= 1 - \theta \frac{b^T y^{k-1} - z^{k-1}}{z^* - z^{k-1}} \\
&\leq 1 - \frac{\theta}{2(1 + \frac{n}{q})} =: 1 - \theta',
\end{aligned} \tag{3.19}$$

where the inequality follows from (3.18). Hence, after k outer iterations we get

$$\begin{aligned}
b^T y^k - z^k &\leq z^* - z^k \\
&\leq (1 - \theta')(z^* - z^{k-1}) \\
&\leq (1 - \theta')^k(z^* - z^0),
\end{aligned}$$

where the second inequality follows from (3.19). This means that

$$b^T y^K - z^k \leq \frac{\epsilon}{4(1 + \frac{n}{q})}$$

certainly holds if

$$(1 - \theta')^K (z^* - z^0) \leq \frac{\epsilon}{4(1 + \frac{n}{q})}.$$

Taking logarithms, this inequality reduces to

$$-K \ln (1 - \theta') \geq \ln \frac{4(1 + \frac{n}{q})(z^* - z^0)}{\epsilon}.$$

Since $-\ln(1 - \theta') \geq \theta'$, this will certainly hold if

$$
\begin{aligned}
K &\geq \frac{1}{\theta'} \ln \frac{4(1 + \frac{n}{q})(z^* - z^0)}{\epsilon} \\
&= \frac{2}{\theta}(1 + \frac{n}{q}) \ln \frac{4(1 + \frac{n}{q})(z^* - z^0)}{\epsilon}.
\end{aligned}
$$

If the Center Algorithm stops with solution y, we have, using (3.18),

$$
\begin{aligned}
z^* - b^T y &\leq z^* - z^k \\
&\leq 2(1 + \frac{n}{q})(b^T y - z^k) \\
&\leq \epsilon,
\end{aligned}
$$

which means that y is an ϵ-optimal solution. This proves the theorem. □

It is well-known that a 2^{-2L}-optimal dual solution can be rounded to an optimal solution for $(\mathcal{LD})$ in $O(n^3)$ arithmetic operations. (See e.g. [120].) Consequently, for this purpose it suffices to take $\epsilon = 2^{-2L}$.

The following lemma is needed to derive an upper bound for the number of inner iterations in each outer iteration. It states that a sufficient decrease in the distance function value can be obtained by taking a step along the Newton direction.

Lemma 3.7 *Let* $\bar{\alpha} := (1 + \sigma)^{-1}$. *Then*

$$\Delta\phi_{\mathrm{D}} := \phi_{\mathrm{D}}(y, z) - \phi_{\mathrm{D}}(y + \bar{\alpha}p, z) \geq \sigma - \ln(1 + \sigma).$$

Proof: Using the interpretation of $\phi_{\mathrm{D}}(y, z)$ as being the logarithmic barrier function for the problem

$$\max\{0 \; : \; \tilde{A}^T y + \tilde{s} = c, \; \tilde{s} \geq 0\},$$

this is an immediate consequence of Lemma 2.6. □

Now we give an upper bound for the total number of inner iterations during an arbitrary outer iteration.

Theorem 3.2 *Each outer iteration requires at most*

$$\frac{11}{3} + 11q\theta \left(\frac{\theta}{1 - \theta} + \frac{2\sqrt{n}}{q + 2\sqrt{n}} \right)$$

inner iterations.

Proof: We denote the lower bound used in an arbitrary outer iteration by $\bar{\bar{z}}$, while the lower bound in the previous outer iteration is denoted by $\bar{z}$. The iterate at the beginning of the outer iteration is denoted by y. Hence y is centered with respect to $y(\bar{z})$ and

$$\bar{\bar{z}} = \bar{z} + \theta(b^T y - \bar{z}). \tag{3.20}$$

For each inner iteration we know, according to Lemma 3.7, that the decrease in the distance function is at least

$$\sigma - \ln(1 + \sigma). \tag{3.21}$$

Since $\sigma > \frac{1}{2}$ in each iteration, and since (3.21) is increasing in σ, it follows that

$$\Delta \phi_D \geq \frac{1}{2} - \ln(1 + \frac{1}{2}) > \frac{1}{11}. \tag{3.22}$$

Let N denote the number of inner iterations. The difference between the distance function value in the current iterate y and the next center $y(\bar{\bar{z}})$ is

$$\phi_D(y, \bar{\bar{z}}) - \phi_D(y(\bar{\bar{z}}), \bar{\bar{z}}).$$

From (3.22) we know that in each inner iteration the distance function value decreases by at least $\frac{1}{11}$. Hence we have

$$\frac{1}{11} N \leq \phi_D(y, \bar{\bar{z}}) - \phi_D(y(\bar{\bar{z}}), \bar{\bar{z}}). \tag{3.23}$$

Let us call the right–hand side of (3.23) $\Phi_D(y, \bar{\bar{z}})$, i.e.

$$\Phi_D(y, \bar{\bar{z}}) \leq \phi_D(y, \bar{\bar{z}}) - \phi_D(y(\bar{\bar{z}}), \bar{\bar{z}}).$$

According to the Mean Value Theorem there is a $\hat{z} \in (\bar{z}, \bar{\bar{z}})$ such that

$$\Phi_D(y, \bar{\bar{z}}) = \Phi_D(y, \bar{z}) + \left. \frac{d\,\Phi_D(y, z)}{d\,z} \right|_{z=\hat{z}} (\bar{\bar{z}} - \bar{z}). \tag{3.24}$$

Let us now look at $\frac{d\,\Phi_D(y,z)}{d\,z}$. We have

$$\frac{d\,\phi_D(y, z)}{d\,z} = \frac{q}{b^T y - z},$$

and, denoting the derivative of $y(z)$ with respect to z by y',

$$\begin{aligned}
\frac{d\,\phi_D(y(z), z)}{d\,z} &= -q \frac{b^T y' - 1}{b^T y(z) - z} + \sum_{i=1}^{n} \frac{a_i^T y'}{s_i(z)} \\
&= -q \frac{b^T y' - 1}{b^T y(z) - z} + \frac{q}{b^T y(z) - z} \sum_{i=1}^{n} x_i(z) a_i^T y' \\
&= \frac{q}{b^T y(z) - z},
\end{aligned}$$

where the last two equations follow from (3.9). So

$$
\begin{aligned}
\left.\frac{d\,\Phi_{\mathrm{D}}(y,z)}{d\,z}\right|_{z=\hat{z}} &= \left. q\left(\frac{1}{b^T y - z} - \frac{1}{b^T y(z) - z}\right)\right|_{z=\hat{z}} \\
&\le q\left(\frac{1}{b^T y - \bar{z}} - \frac{1}{b^T y(\bar{z}) - \bar{z}}\right),
\end{aligned}
$$

where the last inequality follows from the fact that $\bar{z} > \hat{z}$ and from Lemma 3.1. Substituting this into (3.24) gives

$$
\begin{aligned}
\Phi_{\mathrm{D}}(y,\bar{\bar{z}}) &\le \Phi_{\mathrm{D}}(y,\bar{z}) + q\left(\frac{1}{b^T y - \bar{z}} - \frac{1}{b^T y(\bar{z}) - \bar{z}}\right)(\bar{\bar{z}} - \bar{z}) \\
&= \Phi_{\mathrm{D}}(y,\bar{z}) + q\theta\left(\frac{1}{1-\theta} - \frac{b^T y - \bar{z}}{b^T y(\bar{z}) - \bar{z}}\right) \\
&\le \frac{1}{3} + q\theta\left(\frac{1}{1-\theta} - \frac{1}{1+\frac{2\sqrt{n}}{q}}\right) \qquad (3.25) \\
&= \frac{1}{3} + q\theta\left(\frac{\theta}{1-\theta} + \frac{2\sqrt{n}}{q+2\sqrt{n}}\right), \qquad (3.26)
\end{aligned}
$$

where the first equality follows from (3.20) and inequality (3.25) follows because $\Phi_{\mathrm{D}}(y,\bar{z}) \le \frac{1}{3}$ according to Lemma 3.5, and

$$
b^T y(\bar{z}) - \bar{z} \le (1 + \frac{2\sqrt{n}}{q})(b^T y - \bar{z})
$$

according to Lemma 3.6. The theorem follows by combining (3.26) and (3.23):

$$
\begin{aligned}
N &\le 11\Phi_{\mathrm{D}}(y,\bar{\bar{z}}) \\
&\le \frac{11}{3} + 11q\theta\left(\frac{\theta}{1-\theta} + \frac{2\sqrt{n}}{q+2\sqrt{n}}\right)
\end{aligned}
$$

$\square$

Combining Theorems 3.1 and 3.2 we have the following theorem

Theorem 3.3 *After at most*

$$
22(1 + \frac{n}{q})\left(\frac{1}{3\theta} + q\left(\frac{\theta}{1-\theta} + \frac{2\sqrt{n}}{q+2\sqrt{n}}\right)\right)\ln\frac{4(1+\frac{n}{q})(z^* - z^0)}{\epsilon}
$$

Newton iterations, the algorithm ends up with an ϵ-optimal solution. $\square$

This makes clear that to obtain an ϵ-optimal solution the algorithm needs (with $q = \Theta(n)$)

• $O(n\ln\frac{z^*-z^0}{\epsilon})$ Newton iterations for the long–step variant ($0 < \theta < 1$);

- $O(\sqrt{n}\ln\frac{z^*-z^0}{\epsilon})$ Newton iterations for the medium–step variant ($\theta = \frac{\nu}{\sqrt{n}}$, $\nu > 0$).

To obtain an optimal solution we have to take $\epsilon = 2^{-2L}$. For this value of ϵ the iteration bounds become $O(nL)$ and $O(\sqrt{n}L)$ respectively, assuming that $z^* - z^0 \leq 2^{O(L)}$. Note that the best complexity result for the medium–step variant is obtained by taking $q = \Theta(n)$. The given complexity bound for the long–step variant can also be obtained by using $2\sqrt{n} \leq q = \Theta(\sqrt{n})$.

At the end of each sequence of inner iterations we have a dual feasible y such that $\sigma(y,z) \leq 1$. The following lemma shows that a primal feasible solution can also be obtained in this region.

Lemma 3.8 *Let the first n components of $\tilde{x} := \tilde{x}(y,z)$ be denoted by $\hat{x}$. If $\sigma := \sigma(y,z) \leq 1$ then*

$$x := \frac{\hat{x}}{q\tilde{x}_{n+1}}$$

is primal feasible. Moreover,

$$\frac{n-\sigma\sqrt{n}}{q+\sigma\sqrt{q}}(b^T y - z) \leq c^T x - b^T y \leq \frac{n+\sigma\sqrt{n}}{q-\sigma\sqrt{q}}(b^T y - z).$$

Proof: By the definition of $\tilde{x}(y,z)$ (see (3.14) we have

$$\|\tilde{S}\tilde{x}(y,z) - e\| \leq 1.$$

This implies $\tilde{x}(y,z) \geq 0$, so $x = \frac{\hat{x}}{q\tilde{x}_{n+1}} \geq 0$. Moreover, we have $\tilde{A}\tilde{x} = 0$, and since the last q components of $\tilde{x}$ are equal and the last q columns of $\tilde{A}$ are equal to b, this is equivalent to

$$A\hat{x} = q\tilde{x}_{n+1}b.$$

So

$$A\frac{\hat{x}}{q\tilde{x}_{n+1}} = b,$$

which means that

$$x = \frac{\hat{x}}{q\tilde{x}_{n+1}}$$

is primal feasible. Moreover,

$$
\begin{aligned}
\left| s^T \hat{x} - n \right| &= \left| e^T(S\hat{x} - e) \right| \\
&\leq \|e\|\,\|S\hat{x} - e\| \\
&\leq \sigma\sqrt{n},
\end{aligned}
$$

where the last inequality follows because

$$
\begin{aligned}
\sigma^2 &= \|\tilde{S}\tilde{x}(y,z) - e\|^2 \\
&= \|S\hat{x} - e\|^2 + q(\tilde{s}_{n+1}\tilde{x}_{n+1} - 1)^2 \qquad (3.27) \\
&\geq \|S\hat{x} - e\|^2.
\end{aligned}
$$

Consequently,

$$n - \sigma\sqrt{n} \le s^T\hat{x} \le n + \sigma\sqrt{n}.$$

From (3.27) we also derive that

$$q - \sigma\sqrt{q} \le q\tilde{x}_{n+1}\tilde{s}_{n+1} \le q + \sigma\sqrt{q}. \tag{3.28}$$

Hence, since $\tilde{s}_{n+1} = b^T y - z$,

$$\frac{n - \sigma\sqrt{n}}{q + \sigma\sqrt{q}}(b^T y - z) \le \frac{\hat{x}^T s}{q\tilde{x}_{n+1}} \le \frac{n + \sigma\sqrt{n}}{q - \sigma\sqrt{q}}(b^T y - z),$$

which proves the lemma.   □

Based on the quadratic convergence (Lemma 3.4) we can easily analyze the short-step center method. The following lemma shows the effect on σ of shifting the 'objective constraints' with a factor $1 - \theta$, which is equivalent to updating the lower bound.

Lemma 3.9 Let $\sigma := \sigma(y, z)$ and $z^+ := z + \theta(b^T y - z)$. Then

$$\sigma(y, z^+) \le (1 + \theta)\sigma + \theta\sqrt{q}.$$

Proof: First note that s_i, $i = 1, \cdots, n$, does not change after shifting, but the slacks for the objective constraints become $(1 - \theta)s_i$, $i = n+1, \cdots, n+q$. So, if I_q denotes the $(n+q) \times (n+q)$ diagonal matrix obtained from the identity matrix by replacing the first n 1's on the diagonal by 0, then the new slack variable becomes $(I - \theta I_q)\tilde{s}$. By definition

$$\sigma(y, z^+) = \min_{\bar{x}}\left\{\left\|(I - \theta I_q)\tilde{S}\bar{x} - e\right\| \; : \; \tilde{A}\bar{x} = 0\right\}.$$

Take $\bar{x} = \tilde{x}(y, z)$, where $\tilde{x} := \tilde{x}(y, z)$ solves the minimization problem corresponding to $\sigma(y, z)$. Then it follows that

$$\begin{aligned}
\sigma(y, z^+) &\le \left\|\tilde{S}\tilde{x} - e - \theta I_q\tilde{S}\tilde{x}\right\| \\
&\le \left\|\tilde{S}\tilde{x} - e\right\| + \left\|\theta I_q\tilde{S}\tilde{x}\right\| \\
&= \sigma + \theta\left\|I_q\tilde{S}\tilde{x}\right\| \\
&= \sigma + \theta\sqrt{q}\tilde{x}_{n+1}\tilde{s}_{n+1} \\
&\le \sigma + \theta(\sqrt{q} + \sigma) \\
&= (1 + \theta)\sigma + \theta\sqrt{q},
\end{aligned}$$

where the last inequality follows from (3.28). This proves the lemma.   □

As a consequence of the previous lemma and Lemma 3.4 we have the following lemma.

Lemma 3.10 Let $y^+ := y + p$ and $z^+ := z + \theta(b^T y - z)$, where $\theta = \frac{1}{8\sqrt{q}}$. If $\sigma(y, z) \le \frac{1}{2}$ then $\sigma(y^+, z^+) \le \frac{1}{2}$.

Proof: As a consequence of Lemma 3.9 we have

$$
\begin{aligned}
\sigma(y, z^+) &\leq (1+\theta)\sigma(y,z) + \theta\sqrt{q} \\
&\leq (1 + \frac{1}{8\sqrt{q}})\frac{1}{2} + \frac{1}{8\sqrt{q}}\sqrt{q} \\
&\leq \frac{11}{16}.
\end{aligned}
$$

Now using Lemma 3.4, the lemma follows:

$$
\begin{aligned}
\sigma(y^+, z^+) &\leq \sigma(y, z^+)^2 \\
&\leq \frac{121}{256} \\
&< \frac{1}{2}.
\end{aligned}
$$

This proves the lemma. $\square$

Consequently, if we take e.g. $q = \Theta(n)$, then if follows from Theorem 3.1 that the short-step center method requires $O(\sqrt{n}\ln\frac{z^* - z^0}{\epsilon})$ Newton iterations. The same complexity bound is obtained by Renegar [127], but our analysis is much simpler.

3.4 Smooth convex programming

3.4.1 On the monotonicity of the primal and dual objectives along the path of centers

In Section 2.5 we proved the monotonicity of the primal and dual objective values along the path defined by the logarithmic barrier path. Although this path coincides with the path of centers defined by the distance function, this does not necessarily imply monotonicity for the objective values in z (the parameterization of the central path is different).

In Den Hertog et al. [28] we proved that the primal objective $f_0(y(z))$ is increasing and $f_0(y(z)) - z$ is decreasing if z increases. This was done by differentiating the Karush–Kuhn–Tucker conditions and manipulating these equations. In that proof we needed the objective and constraint functions to be convex and twice continuously differentiable.

In this section we will prove a more general result. Assuming only convexity and differentiability of the functions $-f_0(y)$ and $f_i(y), i = 1, \cdots, n$, we prove that the objective value $f_0(y(z))$ increases, and the dual objective and $f_0(y(z)) - z$ decrease, if z increases.

Theorem 3.4 *The primal objective $f_0(y(z))$ is monotonically increasing, the dual objective $f_0(y(z)) - \sum_{i=1}^{n} x_i(z)f_i(y(z))$ and $f_0(y(z)) - z$ are monotonically decreasing if z increases.*

Proof: We first prove the last part of the theorem. Suppose $\bar{z} > z$, such that $f_0(y(z)) > \bar{z}$. Consider the function:

$$\phi_D^d(\tilde{x}, y, z) = \sum_{i=1}^{n+q} \tilde{x}_i f_i(y) + \sum_{i=1}^{n+q} \ln \tilde{x}_i + (n+q).$$

Note that while $\phi_D(y, z)$ can be interpreted as the logarithmic barrier function for the problem

$$\max\{0 : f_i(y) \le 0, \ 1 \le i \le n+q\},$$

we have that $\phi_D^d(\tilde{x}, y, z)$ can be interpreted (up to a constant) as the logarithmic barrier function for its dual problem

$$\min\{-\sum_{i=1}^{n+q} \tilde{x}_i f_i(y) \ : \ \sum_{i=1}^{n+q} \tilde{x}_i \nabla f_i(y) = 0, \ \tilde{x} \ge 0\}.$$

It is easy to verify that a maximizing point for $\phi_D^d(\tilde{x}, y, z)$ has to satisfy (3.5). Hence $(\tilde{x}(z), y(z))$ is the unique maximizer of $\phi_D^d(\tilde{x}, y, z)$. Note that, due to (3.5), the last q coordinates of $\tilde{x}$ are equal to $\tilde{x}_{n+1}$. The first n components of $\tilde{x}$ are denoted as the vector $\hat{x}$. Now, since $(\tilde{x}(z), y(z))$ maximizes $\phi_D^d(\tilde{x}, y, z)$ and $(\tilde{x}(\bar{z}), y(\bar{z}))$ maximizes $\phi_D^d(\tilde{x}, y, \bar{z})$, we have

$$\phi_D^d(\tilde{x}(z), y(z), z) \ge \phi_D^d(\tilde{x}(\bar{z}), y(\bar{z}), z)$$

amd

$$\phi_D^d(\tilde{x}(\bar{z}), y(\bar{z}), \bar{z}) \ge \phi_D^d(\tilde{x}(z), y(z), \bar{z})$$

or equivalently

$$\sum_{i=1}^{n} \hat{x}_i(z) f_i(y(z)) + q \left[z - f_0(y(z))\right] \tilde{x}_{n+1}(z) + \sum_{i=1}^{n+q} \ln \tilde{x}_i(z) + (n+q)$$

$$\ge \sum_{i=1}^{n} \hat{x}_i(\bar{z}) f_i(y(\bar{z})) + q \left[z - f_0(y(\bar{z}))\right] \tilde{x}_{n+1}(\bar{z}) + \sum_{i=1}^{n+q} \ln \tilde{x}_i(\bar{z}) + (n+q),$$

and

$$\sum_{i=1}^{n} \hat{x}_i(\bar{z}) f_i(y(\bar{z})) + q \left[\bar{z} - f_0(y(\bar{z}))\right] \tilde{x}_{n+1}(\bar{z}) + \sum_{i=1}^{n+q} \ln \tilde{x}_i(\bar{z}) + (n+q)$$

$$\ge \sum_{i=1}^{n} \hat{x}_i(z) f_i(y(z)) + q \left[\bar{z} - f_0(y(z))\right] \tilde{x}_{n+1}(z) + \sum_{i=1}^{n+q} \ln \tilde{x}_i(z) + (n+q).$$

Adding the two inequalities gives

$$q \left[z\tilde{x}_{n+1}(z) + \bar{z}\tilde{x}_{n+1}(\bar{z})\right] \ge q \left[z\tilde{x}_{n+1}(\bar{z}) + \bar{z}\tilde{x}_{n+1}(z)\right],$$

or equivalently,

$$(z - \bar{z})(\tilde{x}_{n+1}(z) - \tilde{x}_{n+1}(\bar{z})) \ge 0.$$

This means that

$$\tilde{x}_{n+1}(z) \le \tilde{x}_{n+1}(\bar{z}). \tag{3.29}$$

Now from (3.5) we have

$$\tilde{x}_{n+1}(z) = \frac{1}{-f_{n+1}(y(z))} = \frac{1}{f_0(y(z)) - z},$$

and a similar expression for $\tilde{x}_{n+1}(\bar{z})$. Substituting this into (3.29) gives that $f_0(y(z)) - z$ monotonically decreases if z decreases. Finally, note that $y(z) = y(\mu)$, for $\mu = \frac{f_0(y(z)) - z}{q}$, and $y(\bar{z}) = y(\bar{\mu})$, for $\bar{\mu} = \frac{f_0(y(\bar{z})) - \bar{z}}{q}$. Since $\bar{\mu} \leq \mu$, it follows from Theorem 2.8 that $f_0(y(z)) - \sum_{i=1}^{n} x_i(z) f_i(y(z))$ monotonically decreases and $f_0(y(z))$ monotonically increases if z increases.

There is also a direct proof for the monotonicity of $f_0(y(z))$. Since $y(z)$ minimizes $\phi_D(y, z)$ and $y(\bar{z})$ minimizes $\phi_D(y, \bar{z})$ we obviously have

$$\phi_D(y(z), z) \leq \phi_D(y(\bar{z}), z)$$

$$\phi_D(y(\bar{z}), \bar{z}) \leq \phi_D(y(z), \bar{z})$$

or equivalently

$$-q \ln(f_0(y(z)) - z) - \sum_{i=1}^{n} \ln(-f_i(y(z))) \leq$$
$$-q \ln(f_0(y(\bar{z})) - z) - \sum_{i=1}^{n} \ln(-f_i(y(\bar{z}))),$$

and

$$-q \ln(f_0(y(\bar{z})) - \bar{z}) - \sum_{i=1}^{n} \ln(-f_i(y(\bar{z}))) \leq$$
$$-q \ln(f_0(y(z)) - \bar{z}) - \sum_{i=1}^{n} \ln(-f_i(y(z))).$$

Adding the two inequalities gives

$$-q \ln(f_0(y(z)) - z) - q \ln(f_0(y(\bar{z})) - \bar{z}) \leq -q \ln(f_0(y(\bar{z})) - z) - q \ln(f_0(y(z)) - \bar{z}),$$

or equivalently, by taking the exponential

$$-(f_0(y(z)) - z)(f_0(y(\bar{z})) - \bar{z}) \leq -(f_0(y(\bar{z})) - z)(f_0(y(z)) - \bar{z}),$$

which reduces to

$$(z - \bar{z})(f_0(y(\bar{z})) - f_0(y(z))) \leq 0.$$

Since $\bar{z} > z$ this means that $f_0(y(z)) \leq f_0(y(\bar{z}))$. $\qquad \square$

The following lemma gives an upper bound for the gap $z^* - z$.

Lemma 3.11 *One has*

$$z^* - z \leq (1 + \frac{n}{q})(f_0(y(z)) - z).$$

Proof: The exact center $y(z)$ minimizes the distance function $\phi_D(y, z)$. The necessary and sufficient conditions for these minima are (3.2). From these conditions we derive that $(x(z), y(z))$ is dual feasible. Moreover, using that the dual objective value is always greater than or equal to the optimal value, i.e.

$$z^* \leq f_0(y(z)) - \sum_{i=1}^{n} x_i(z) f_i(y(z)),$$

it follows that

$$z^* - f_0(y(z)) \leq -\sum_{i=1}^{n} x_i(z) f_i(y(z)) = \frac{n}{q}(f_0(y(z)) - z).$$

Consequently,

$$(z^* - z) - (f_0(y(z)) - z) \leq \frac{n}{q}(f_0(y(z)) - z).$$

This means

$$z^* - z \leq (1 + \frac{n}{q})(f_0(y(z)) - z).$$

This proves the lemma. $\square$

3.4.2   Properties near the centers

For the analysis we again assume that the functions $-f_0(y)$ and $f_i(y)$, $0 \leq i \leq n$, are convex with continuous first and second order derivatives in $\mathcal{F}$, and moreover that $\phi_D(y, z)$ is κ–self–concordant. This self–concordance property was introduced and discussed in Section 2.5.2. The distance functions of the classes of problems which were mentioned in that section also satisfy this condition (see also Appendix A). The strict convexity of $\phi_D(y, z)$ follows from the boundedness of $\mathcal{F}$ and the self–concordance of $\phi_D(y, z)$ (see [119] and [68]).

Now we will prove some lemmas for approximately centered points. As indicated in Section 3.1, we can use the results obtained for the logarithmic barrier method in Section 2.5, since $y(z)$ is the maximum solution for $\phi_D(y, z)$, which can be viewed as the logarithmic barrier function for the problem

$$\max\{0 \; : \; f_i(y) \leq 0, \; 1 \leq i \leq n + q\}.$$

Lemma 3.12 *Let $d \in \mathbb{R}^m$. If $\|d\|_{H(y,z)} < \frac{1}{\kappa}$ then $y + d \in \mathcal{F}_z^0 \subseteq \mathcal{F}^0$.*

Proof: Since $\phi_D(y, z)$ is κ–self–concordant, this is an immediate consequence of Lemma 2.20. $\square$

Lemma 3.13 *Let $y^+ := y + p$. If $\|p\|_H := \|p(y, z)\|_{H(y,z)} < \frac{1}{\kappa}$ then $y^+ \in \mathcal{F}_z^0$, and*

$$\|p(y^+, z)\|_{H(y^+,z)} \leq \frac{\kappa}{(1 - \kappa\|p\|_H)^2} \|p\|_H^2.$$

Proof: Since $\phi_D(y, z)$ is κ–self-concordant, this is an immediate consequence of Lemma 2.21. □

Again, as a consequence of Lemma 3.13, for $\|p\|_H < \frac{3-\sqrt{5}}{2\kappa}$ we have that

$$\|p(y^+, z)\|_{H(y^+, z)} < \|p\|_H,$$

and hence convergence of Newton's method. For $\|p\|_H \leq \frac{1}{3\kappa}$ the lemma gives

$$\|p(y^+, z)\|_{H(y^+, z)} \leq \frac{9}{4}\kappa\|p\|_H^2. \tag{3.30}$$

Lemma 3.14 *If* $\|p\|_H \leq \frac{1}{3\kappa}$ *then*

$$\phi_D(y, z) - \phi_D(y(z), z) \leq \frac{\|p\|_H^2}{1 - \left(\frac{9}{4}\kappa\|p\|_H\right)^2}.$$

Proof: Using the interpretation of $\phi_D(y, z)$ as being a logarithmic barrier function, this lemma is an immediate consequence of Lemma 2.22. □

Lemma 3.15 *If* $\|p\|_H := \|p(y, z)\|_{H(y, z)} \leq \frac{1}{3\kappa}$ *and* $q \geq \frac{\sqrt{n}}{\kappa}$, *then*

$$f_0(y(z)) - z \leq \left(1 + \frac{2\sqrt{n}}{q}\frac{\|p\|_H}{1 - \frac{9}{4}\kappa\|p\|_H}\right)(f_0(y) - z).$$

Proof: Since $p = -H^{-1}g$ we have $p^T g = -\|p\|_H^2$. On the other hand

$$-p^T g = p^T \left(\frac{q \nabla f_0(y)}{f_0(y) - z} - \sum_{i=1}^n \frac{\nabla f_i(y)}{-f_i(y)}\right).$$

So we have

$$p^T \nabla f_0(y) = \frac{f_0(y) - z}{q}\left(\|p\|_H^2 + p^T \sum_{i=1}^n \frac{\nabla f_i(y)}{-f_i(y)}\right). \tag{3.31}$$

Now let J denote the matrix whose columns are $\frac{\nabla f_i(y)}{-f_i(y)}$. Now it is easy to see that

$$\left|p^T \sum_{i=1}^n \frac{\nabla f_i(y)}{-f_i(y)}\right| = |p^T J e|$$

$$\leq \|J^T p\|\,\|e\|$$

$$\leq \sqrt{n}\|p\|_H.$$

The last inequality follows from

$$\|J^T p\|^2 = p^T J J^T p$$

$$= p^T \sum_{i=1}^n \frac{\nabla f_i(y)\nabla f_i(y)^T}{f_i(y)^2}p$$

$$\leq p^T H p$$

$$= \|p\|_H^2.$$

(The inequality holds since from (3.6) we have

$$\frac{\nabla f_i(y)\nabla f_i(y)^T}{f_i(y)^2} \preceq H.)$$

Using this in (3.31) gives

$$p^T \nabla f_0(y) \leq \frac{f_0(y) - z}{q}\left(\|p\|_H^2 + \sqrt{n}\|p\|_H\right)$$

$$\leq 2\|p\|_H \frac{f_0(y) - z}{q}\sqrt{n}. \tag{3.32}$$

Let $y^0 := y$ and let $y^1, y^2, \cdots$ denote the sequence of points obtained by repeating Newton steps, starting at y^0. According to Lemma 3.13 we have

$$\|p(y^k, z)\|_{H(y^k,z)} \leq \frac{9}{4}\kappa\|p(y^{k-1}, z)\|_{H(y^{k-1},z)}^2$$

$$\vdots$$

$$\leq \left(\frac{9}{4}\kappa\right)^{2^k - 1}\|p(y^0, z)\|_{H(y^0,z)}^{2^k}. \tag{3.33}$$

This means that $\|p(y^k, z)\|_{H(y^k,z)} \to 0$ if $k \to \infty$. Since all y^k lie in a bounded region and $\|p(y, z)\|_{H(y,z)} = 0$ if and only if $y = y(z)$, we have that all limit points of the sequence are $y(z)$. Hence, the sequence of Newton iterates converges to $y(z)$. Therefore, we may write

$$\begin{aligned}
f_0(y^{k+1}) - z &= f_0(y^k) - z + f_0(y^{k+1}) - f_0(y^k) \\
&\leq f_0(y^k) - z + \nabla f_0(y^k)^T p(y^k, z) \\
&\leq \left(1 + 2\|p(y^k, z)\|_{H(y^k,z)}\frac{\sqrt{n}}{q}\right)(f_0(y^k) - z) \\
&\leq \left(1 + 2\|p(y^k, z)\|_{H(y^k,z)}\frac{\sqrt{n}}{q}\right) \times \\
&\qquad \left(1 + 2\|p(y^{k-1}, z)\|_{H(y^{k-1},z)}\frac{\sqrt{n}}{q}\right)(f_0(y^{k-1}) - z)
\end{aligned}$$

$$\vdots$$

$$\leq (f_0(y^0) - z)\prod_{i=0}^{k}\left(1 + 2\|p(y^i, z)\|_{H(y^i,z)}\frac{\sqrt{n}}{q}\right)$$

$$\leq (f_0(y^0) - z)\prod_{i=0}^{k}\left(1 + 2\left(\frac{9}{4}\kappa\right)^{2^i - 1}\|p\|_H^{2^i}\frac{\sqrt{n}}{q}\right),$$

where the first inequality follows from the convexity of $-f_0(y)$, and the second inequality from (3.32) and the last inequality from (3.33). In particular we have:

$$f_0(y(z)) - z \leq (f_0(y^0) - z)\prod_{i=0}^{\infty}\left(1 + 2\left(\frac{9}{4}\kappa\right)^{2^i - 1}\|p\|_H^{2^i}\frac{\sqrt{n}}{q}\right)$$

$$\leq \left(1 + \frac{2\sqrt{n}}{q}\frac{\|p\|_H}{1 - \frac{9}{4}\kappa\|p\|_H}\right)(f_0(y^0) - z),$$

where the last inequality follows from Lemma B.4 and because we assumed $q \geq \frac{\sqrt{n}}{\kappa}$.

$\square$

3.4.3 Complexity analysis

Based on the lemmas in the previous section, we will give upper bounds for the total number of outer iterations and inner iterations needed by the Center Algorithm stated in Section 3.1. First we will deal with the long– and medium–step variant, and finally with the short–step variant. We set $\tau = \frac{1}{3\kappa}$ in the Center Algorithm.

Theorem 3.5 *After at most*

$$\frac{4}{\theta}(1 + \frac{n}{q}) \ln \frac{4(1 + \frac{n}{q})(z^* - z^0)}{\epsilon}$$

outer iterations, the Center Algorithm ends up with an ϵ-optimal solution for (CP).

Proof: The proof is similar to the proof of Theorem 3.1. Let z^k be the lower bound in the k–th outer iteration, and y^k the iterate at the end of k outer iterations. Using Lemmas 3.11 and 3.15 and the fact that $q \geq \frac{\sqrt{n}}{\kappa}$ successively, we obtain

$$
\begin{aligned}
z^* - z^{k-1} &\leq (1 + \frac{n}{q})(f_0(y(z^{k-1})) - z^{k-1}) \\
&\leq (1 + \frac{n}{q})\left(1 + \frac{2\sqrt{n}}{q}\frac{\frac{1}{3\kappa}}{1 - \frac{9}{4\kappa}\cdot\frac{1}{3\kappa}}\right)(f_0(y^{k-1}) - z^{k-1}) \\
&\leq 4(1 + \frac{n}{q})(f_0(y^{k-1}) - z^{k-1}). \quad (3.34)
\end{aligned}
$$

Moreover,

$$
\begin{aligned}
\frac{z^* - z^k}{z^* - z^{k-1}} &= \frac{z^* - (z^{k-1} + \theta(f_0(y^{k-1}) - z^{k-1}))}{z^* - z^{k-1}} \\
&= 1 - \theta\frac{f_0(y^{k-1}) - z^{k-1}}{z^* - z^{k-1}} \\
&\leq 1 - \frac{\theta}{4(1 + \frac{n}{q})} =: 1 - \theta', \quad (3.35)
\end{aligned}
$$

where the inequality follows from (3.34). Hence, after k outer iterations we get

$$
\begin{aligned}
f_0(y^k) - z^k &\leq z^* - z^k \\
&\leq (1 - \theta')(z^* - z^{k-1}) \\
&\leq (1 - \theta')^k (z^* - z^0),
\end{aligned}
$$

where the second inequality follows from (3.35). This means that

$$f_0(y^K) - z^k \leq \frac{\epsilon}{4(1 + \frac{n}{q})}$$

certainly holds if

$$(1 - \theta')^K (z^* - z^0) \leq \frac{\epsilon}{4(1 + \frac{n}{q})}.$$

Taking logarithms, this inequality reduces to

$$-K \ln (1 - \theta') \geq \ln \frac{4(1 + \frac{n}{q})(z^* - z^0)}{\epsilon}.$$

Since $- \ln(1 - \theta') \geq \theta'$, this will certainly hold if

$$
\begin{aligned}
K &\geq \frac{1}{\theta'} \ln \frac{4(1 + \frac{n}{q})(z^* - z^0)}{\epsilon} \\
&= \frac{4}{\theta}(1 + \frac{n}{q}) \ln \frac{4(1 + \frac{n}{q})(z^* - z^0)}{\epsilon}.
\end{aligned}
$$

If the Center Algorithm stops with solution y, we have, using (3.34),

$$
\begin{aligned}
z^* - f_0(y) &\leq z^* - z^k \\
&\leq 4(1 + \frac{n}{q})(f_0(y) - z^k) \\
&\leq \epsilon,
\end{aligned}
$$

which means that y is an ϵ–optimal solution. This proves the theorem. $\square$

The following lemma states that a sufficient decrease in the distance function can be obtained by taking a step along the Newton direction.

Lemma 3.16 *Let* $\bar{\alpha} := \frac{1}{1 + \kappa \|p\|_H}$ *then*

$$\phi_D(y, z) - \phi_D(y + \bar{\alpha}p, z) \geq \frac{1}{\kappa^2}(\kappa \|p\|_H - \ln(1 + \kappa \|p\|_H)).$$

Proof: Since $\phi_D(y, z)$ is κ–self–concordant, this lemma is an immediate consequence of Lemma 2.24. $\square$

Now we give an upper bound for the total number of inner iterations during an arbitrary outer iteration.

Theorem 3.6 *The total number of inner iterations during an arbitrary outer iteration is at most*

$$\frac{22}{3} + 22q\kappa^2 \theta \left(\frac{\theta}{1 - \theta} + \frac{3\sqrt{n}}{q\kappa + 3\sqrt{n}} \right).$$

Proof: We denote the used lower bound in an arbitrary outer iteration by $\bar{z}$, while the lower bound in the previous outer iteration is denoted by $\bar{z}$. The iterate at the beginning of the outer iteration is denoted by y. Hence y is centered with respect to $y(\bar{z})$ and

$$\bar{\bar{z}} = \bar{z} + \theta(f_0(y) - \bar{z}). \tag{3.36}$$

Because of Lemma 3.16 we have that the decrease in the distance function is at least

$$\frac{1}{\kappa^2}(\kappa\|p\|_H - \ln(1 + \kappa\|p\|_H)).$$

Since this expression is increasing in $\|p\|_H$ and because $\|p\|_H \geq \frac{1}{3\kappa}$ during each inner iteration we obtain that the decrease in each iteration is at least

$$\frac{1}{\kappa^2}\left(\frac{1}{3} - \ln(1 + \frac{1}{3})\right) > \frac{1}{22\kappa^2}.$$

The difference is distance function value in the current iterate and the next center $\bar{\bar{z}}$ is

$$\phi_D(y, \bar{\bar{z}}) - \phi_D(y(\bar{\bar{z}}), \bar{\bar{z}}).$$

Hence we may write

$$\frac{N}{22\kappa^2} \leq \phi_D(y, \bar{\bar{z}}) - \phi_D(y(\bar{\bar{z}}), \bar{\bar{z}}), \tag{3.37}$$

where N denotes the number of inner iterations. Let us call the right-hand side of (3.37) $\Phi_D(y, \bar{\bar{z}})$, i.e.

$$\Phi_D(y, \bar{\bar{z}}) = \phi_D(y, \bar{\bar{z}}) - \phi_D(y(\bar{\bar{z}}), \bar{\bar{z}}).$$

According to the Mean Value Theorem there is a $\hat{z} \in (\bar{z}, \bar{\bar{z}})$ such that

$$\Phi_D(y, \bar{\bar{z}}) = \Phi_D(y, \bar{z}) + \left.\frac{d\,\Phi_D(y, z)}{d\,z}\right|_{z=\hat{z}} (\bar{\bar{z}} - \bar{z}). \tag{3.38}$$

Let us now look at $\frac{d\,\Phi_D(y,z)}{d\,z}$. We have

$$\frac{d\,\phi_D(y, z)}{d\,z} = \frac{q}{f_0(y) - z},$$

and

$$\begin{aligned}
\frac{d\,\phi_D(y(z), z)}{d\,z} &= -q\frac{\nabla f_0(y(z))^T y' - 1}{f_0(y(z)) - z} + \sum_{i=1}^{n}\frac{\nabla f_i(y(z))^T y'}{-f_i(y(z))} \\
&= -q\frac{\nabla f_0(y(z))^T y' - 1}{f_0(y(z)) - z} + \frac{q}{f_0(y(z)) - z}\sum_{i=1}^{n} x_i(z)\nabla f_i(y(z))^T y' \\
&= \frac{q}{f_0(y(z)) - z},
\end{aligned}$$

where the last two equations follow from (3.2). So

$$\begin{aligned}
\left.\frac{d\,\Phi_D(y, z)}{d\,z}\right|_{z=\hat{z}} &= q\left.\left(\frac{1}{f_0(y) - z} - \frac{1}{f_0(y(z)) - z}\right)\right|_{z=\hat{z}} \\
&\leq q\left(\frac{1}{f_0(y) - \bar{\bar{z}}} - \frac{1}{f_0(y(\bar{z})) - \bar{z}}\right),
\end{aligned}$$

where the last inequality follows from the fact that $\bar{z} > \hat{z}$ and from Theorem 3.4. Substituting this into (3.38) gives

$$
\begin{aligned}
\Phi_{\mathrm{D}}(y,\bar{\bar{z}}) &\leq \Phi_{\mathrm{D}}(y,\bar{z}) + q\left(\frac{1}{f_0(y)-\bar{\bar{z}}} - \frac{1}{f_0(y(\bar{z}))-\bar{z}}\right)(\bar{\bar{z}}-\bar{z}) \\
&= \Phi_{\mathrm{D}}(y,\bar{z}) + q\theta\left(\frac{1}{1-\theta} - \frac{f_0(y)-\bar{z}}{f_0(y(\bar{z}))-\bar{z}}\right) \\
&\leq \frac{1}{3\kappa^2} + q\theta\left(\frac{1}{1-\theta} - \frac{1}{1+\frac{3\sqrt{n}}{q\kappa}}\right) \\
&= \frac{1}{3\kappa^2} + q\theta\left(\frac{\theta}{1-\theta} + \frac{3\sqrt{n}}{q\kappa+3\sqrt{n}}\right),
\end{aligned}
\tag{3.39}
$$

where the first equality follows from (3.36) and the second inequality follows because $\Phi_{\mathrm{D}}(y,\bar{z}) \leq \frac{1}{3\kappa^2}$ according to Lemma 3.14, and

$$
f_0(y(\bar{z})) - \bar{z} \leq \left(1+\frac{3\sqrt{n}}{q\kappa}\right)(f_0(y)-\bar{z})
$$

according to Lemma 3.15. The theorem follows by combining (3.37) and (3.39):

$$
\begin{aligned}
N &\leq 22\kappa^2\Phi_{\mathrm{D}}(y,\bar{\bar{z}}) \\
&= 22\kappa^2\left(\phi_{\mathrm{D}}(y,\bar{\bar{z}}) - \phi_{\mathrm{D}}(y(\bar{\bar{z}}),\bar{\bar{z}})\right). \\
&\leq \frac{22}{3} + 22q\kappa^2\theta\left(\frac{\theta}{1-\theta} + \frac{3\sqrt{n}}{q\kappa+3\sqrt{n}}\right).
\end{aligned}
$$

□

Combining Theorems 3.5 and 3.6, the total number of iterations turns out to be given by the following theorem.

Theorem 3.7 *An upper bound for the total number of Newton iterations is given by*

$$
88\left(1+\frac{n}{q}\right)\left(\frac{1}{3\theta} + q\kappa^2\left(\frac{\theta}{1-\theta} + \frac{3\sqrt{n}}{q\kappa+3\sqrt{n}}\right)\right)\ln\frac{4(1+\frac{n}{q})(z^*-z^0)}{\epsilon}.
$$

□

This makes clear that to obtain an ϵ-optimal solution and setting $q = \Theta(n)$, the algorithm needs

- $O(\kappa^2 n \ln\frac{z^*-z^0}{\epsilon})$ Newton iterations for the long–step variant $(0 < \theta < 1)$;

- $O(\kappa^2\sqrt{n}\ln\frac{z^*-z^0}{\epsilon})$ Newton iterations for the medium–step variant $(\theta = \frac{\nu}{\sqrt{n}}, \nu > 0)$;

- $O(\kappa\sqrt{n}\ln\frac{z^*-z^0}{\epsilon})$ Newton iterations for $\theta = \frac{\nu}{\kappa\sqrt{n}}, \nu > 0$.

We want to point out that a complexity analysis can easily be given for the short–step path–following method using some of the lemmas given above. Short–step path–following methods start at a nearly centered iterate and after the lower bound z is increased by a small factor, a unit Newton step is taken. The updating factor is sufficiently small, such that the new iterate is again nearly centered with respect to the new center.

The following lemma will be needed for analyzing such a short–step path–following method. It states how the distance changes if z is increased.

Lemma 3.17 *Let $z^+ := z + \theta(f_0(y) - z)$. Then*

$$\|p(y, z^+)\|_{H(y,z^+)} \le \|p\|_H + \frac{\theta}{1 - \theta}\sqrt{q}.$$

Proof: Let us first introduce the notation g^+, p^+ and H^+ for $g(y, z^+)$, $p(y, z^+)$ and $H(y, z^+)$, respectively. Note that using (3.6) we have

$$
\begin{aligned}
H(y, z^+) &= -q\frac{\nabla^2 f_0(y)}{f_0(y) - z^+} + q\frac{\nabla f_0(y)\nabla f_0(y)^T}{(f_0(y) - z^+)^2} + \\
&\quad \sum_{i=1}^n \frac{\nabla^2 f_i(y)}{-f_i(y)} + \sum_{i=1}^n \frac{\nabla f_i(y)\nabla f_i(y)^T}{f_i(y)^2} \\
&= H(y, z) - \frac{q\theta}{1 - \theta}\frac{\nabla^2 f_0(y)}{f_0(y) - z} + q\left(\frac{1}{(1 - \theta)^2} - 1\right)\frac{\nabla f_0(y)\nabla f_0(y)^T}{(f_0(y) - z)^2}.
\end{aligned}
$$

Consequently, from Lemma B.3 it follows that $H(y, z^+)^{-1} \preceq H(y, z)^{-1}$. The distance after increasing z becomes:

$$
\begin{aligned}
\|p^+\|_{H^+} &= \sqrt{(g^+)^T (H^+)^{-1} g^+} \\
&\le \sqrt{(g^+)^T H^{-1} g^+} \\
&= \|g^+\|_{H^{-1}} \\
&= \left\|-q\frac{\nabla f_0(y)}{f_0(y) - z^+} + \sum_{i=1}^n \frac{\nabla f_i(y)}{-f_i(y)}\right\|_{H^{-1}} \\
&= \left\|-\frac{q}{1 - \theta}\frac{\nabla f_0(y)}{f_0(y) - z} + \sum_{i=1}^n \frac{\nabla f_i(y)}{-f_i(y)}\right\|_{H^{-1}} \\
&\le \left\|-q\frac{\nabla f_0(y)}{f_0(y) - z} + \sum_{i=1}^n \frac{\nabla f_i(y)}{-f_i(y)}\right\|_{H^{-1}} + \frac{\theta}{1 - \theta}\left\|q\frac{\nabla f_0(y)}{f_0(y) - z}\right\|_{H^{-1}} \\
&= \|p\|_H + \frac{\theta}{1 - \theta}\left\|q\frac{\nabla f_0(y)}{f_0(y) - z}\right\|_{H^{-1}} \qquad (3.40)
\end{aligned}
$$

Let us continue by evaluating $\left\|q\frac{\nabla f_0(y)}{f_0(y)-z}\right\|_{H^{-1}}^2$. Let J be the $m \times q$ matrix whose columns are all $\frac{\nabla f_0(y)}{f_0(y)-z}$. Then we have

$$q\frac{\nabla f_0(y)}{f_0(y) - z} = Je.$$

So, using the explicit expression for H (3.6), we get

$$\left\| q \frac{\nabla f_0(y)}{f_0(y) - z} \right\|_{H^{-1}}^2 = e^T J^T \left(-q \frac{\nabla^2 f_0(y)}{f_0(y) - z} + \sum_{i=1}^{n} \frac{\nabla^2 f_i(y)}{f_i(y)} + \right.$$
$$\left. \sum_{i=1}^{n} \frac{\nabla f_i(y) \nabla f_i(y)^T}{f_i(y)^2} + JJ^T \right)^{-1} Je. \qquad (3.41)$$

Because $-f_0(y)$ and $f_i(y)$, $i = 1, \cdots, n$, are convex, it follows from Lemma B.3 that the eigenvalues of

$$J^T \left(-q \frac{\nabla^2 f_0(y)}{f_0(y) - z} + \sum_{i=1}^{n} \frac{\nabla^2 f_i(y)}{f_i(y)} + \sum_{i=1}^{n} \frac{\nabla f_i(y) \nabla f_i(y)^T}{f_i(y)^2} + JJ^T \right)^{-1} J$$

are all smaller than or equal to one. Consequently, from (3.41) we have

$$\left\| q \frac{\nabla f_0(y)}{f_0(y) - z} \right\|_{H^{-1}}^2 \leq q.$$

Substituting this into (3.40) yields the lemma.   □

Lemma 3.18 *Let* $y^+ := y + p$ *and* $z^+ := z + \theta(f_0(y) - z)$, *where* $\theta = \frac{1}{22\kappa\sqrt{q}}$. *If*

$$\|p(y, z)\|_{H(y,z)} \leq \frac{1}{3\kappa},$$

then

$$\|p(y^+, z^+)\|_{H(y^+, z^+)} \leq \frac{1}{3\kappa}.$$

Proof: Using Lemmas 3.13 and Lemma 3.17 we have

$$\|p(y^+, z^+)\|_{H(y^+, z^+)} \leq \frac{9}{4} \kappa \|p(y, z^+)\|_{H(y, z^+)}^2$$
$$\leq \frac{9}{4} \kappa \left(\frac{1}{3\kappa} + \frac{1}{1 - \frac{1}{22\kappa\sqrt{q}}} \frac{1}{22\kappa} \right)^2$$
$$< \frac{1}{3\kappa}.$$

This proves the lemma.   □

This shows that if θ is sufficiently small, then one unit Newton step is sufficient to reach the vicinity of $y(z^+)$. With the help of Theorem 3.5 it is easy to see that this short–step algorithm requires $O(\kappa\sqrt{n} \ln \frac{z^* - z^0}{\epsilon})$ Newton iterations, using $q = \Theta(n)$.

3.5 Miscellaneous remarks

Further remarks on the center method

Note that the distance function $\phi_D(y, z)$ for linear programming is self–concordant with $\kappa = 1$. The complexity results for linear programming (Section 3.3) and convex programming (Section 3.4) with $\kappa = 1$ are exactly the same. Comparing the complexity results for the logarithmic barrier method and the center method we observe that they are comparable (instead of $n\mu_0$ in the complexity bounds for the logarithmic barrier method, we have $z^* - z^0$ in the corresponding results for the center method, but both expressions indicate the initial gap).

Again we emphasize that the upper bound for the long–step method can be very pessimistic, because of the line searches involved in the inner iterations and because we normally have that $\|p\|_H \gg \tau$: both reasons lead to greater reductions in the distance function value than used in the analysis. Moreover, note that the iteration bound derived for the short–step method is exact. This explains the fact that a large reduction factor gives a worse bound than a small reduction factor, while one would expect the contrary.

A disadvantage of the center method compared with the logarithmic barrier method is that the updating of the lower bound is always restricted: the lower bound must not exceed the objective value in the current iterate. In the logarithmic barrier method there is no such restriction. This is one of the reasons why the center method has always attracted less attention in the literature then the logarithmic barrier method.

There may be some convex programming problems which satisfy the Relative Lipschitz Condition but do not satisfy the self–concordance condition (this only happens if one of the functions involved is not three times continuously differentiable, see the Appendix A). For these problems the analysis given in [28] is still valid.

Relaxing the initial centering condition

Finally we note that the 'centering assumption' $\|p(y^0, z^0)\|_{H(y^0, z^0)} \leq \tau$ can be alleviated to

$$\phi_D(y^0, z^0) - \phi_D(y(z^0), z^0) \leq O(\sqrt{n} \ln \frac{z^* - z^0}{\epsilon})$$

for the medium–step version, and to

$$\phi_D(y^0, z^0) - \phi_D(y(z^0), z^0) \leq O(n \ln \frac{z^* - z^0}{\epsilon}),$$

for the long–step version. This follows easily from Lemma 3.16: in each inner iteration the distance function value decreases by at least $\frac{1}{22\kappa^2}$, hence after respectively

$$O(\kappa^2 \sqrt{n} \ln \frac{z^* - z^0}{\epsilon})$$

and

$$O(\kappa^2 n \ln \frac{z^* - z^0}{\epsilon})$$

iterations we have found an iterate in the neighborhood of $y(z^0)$.

Comparison with other papers

Short–step center methods for linear programming were analyzed by Renegar [127] and Vaidya [149]. Our analysis for such short–step center methods is much simpler and gives the same complexity result.

It is interesting to compare the σ–measure with other measures used in the literature. Renegar [127] showed that in each iteration

$$\|\tilde{S}(z)^{-1}\tilde{s} - e\| \le \frac{1}{46},$$

where $\tilde{s}$ and $\tilde{s}(z)$ are as defined in Section 3.3. Note that this distance measure is not computable since we do not know the center $y(z)$. This measure is related to our σ–measure as follows:

$$\|\tilde{S}(z)^{-1}\tilde{s} - e\| = \|\tilde{S}\tilde{x}(z) - e\| \le \sigma(y, z),$$

where the last inequality follows from the definition of $\sigma(y, z)$ (3.13). Vaidya [149] used $\phi_D(y, z) - \phi_D(y(z), z)$ as a distance measure, which is also not computable (the exact center $y(z)$ is unknown).

Jarre [66] and Mehrotra and Sun [99] analyzed short–step center methods for some classes of convex programming problems. In [66] the $O((1 + M^2)\sqrt{n}\ln\frac{z^*-z^0}{\epsilon})$ complexity bound was obtained for $\theta = \frac{1}{200\sqrt{n}(1+M^2)}$, where M is the so–called Relative Lipschitz constant; see Section 2.5.2. Note that the updating factor θ is dependent on the (possibly unknown) Relative Lipschitz constant M. Note that all these methods only deal with short–step methods, which are unattractive in practice. To accelerate his method, Jarre [66] proposed a (higher order) extrapolation scheme.

Multi–objective programming

The ideas used in the center method can nicely be generalized for multi–objective programming problems. Instead of one objective we now have more, say r, objectives. We will sketch this generalization for the linear programming problem; for nonlinear programming similar arguments can be used.

Suppose there are r (conflicting) objective vectors:

$$b_1, b_2, \cdots, b_r,$$

and the feasible region is again given by the constraints $A^T y \le c$. We introduce a generalized version of the distance function:

$$-q\sum_{i=1}^{r}\ln(b_i^T y - z_i) - \sum_{i=1}^{n}\ln(c_i - a_i^T y).$$

The unique minimum of this function for fixed values of z_i, $i = 1, \cdots, r$, is the analytic center of the following polyhedron

$$\mathcal{F}_z = \{ y \ : \ A^T y \leq c, \ \underbrace{b_i^T y \geq z_i,}_{\text{q times}} \ i = 1, \cdots, r \}.$$

After (approximately) calculating the center of $\mathcal{F}_z$ we can shift the objective constraints:

$$z_i^+ := z_i + \theta_i (b_i^T y - z_i), \ i = 1, \cdots, r,$$

where $0 < \theta_i \leq 1$.

Morin and Trafalis [114] showed that using exact centers $(y = y(z))$ and $\theta_i = 1$, $i = 1, \cdots, r$, this process converges to a point on an efficient facet. (Loosely speaking an efficient facet is a facet where at least one of the objective functions is optimal.) A disadvantage of taking $\theta_i = 1$ is that the current iterate can not be used for the next stage. Generalizing some lemmas given in Section 3.3 it should be possible to analyze the process for $\theta_i < 1$ while using approximate centers. This will also allow us to use different values θ_i or to change these values during the process according to the preference of the user. We expect that similar complexity bounds as for the single objective linear programming problem can be proved.

Chapter 4

Reducing the complexity for LP

In this chapter we will discuss some ways to cut down the number of arithmetic operations per iteration for the logarithmic barrier method for linear programming, analyzed in Section 2.3. The analysis given in this chapter can easily be extended to the center method, because of the close relationship between the logarithmic barrier and the center method as explained in Section 3.3.

Essentially, in each iteration of these methods the $m \times m$ normal matrix $AS^{-2}A^T$ has to be computed and inverted. In the first section we show that the overall complexity can be reduced by a factor $\sqrt{n}$ by using approximations for S and rank–one updates in the computation of the inverse. In the second section we analyze a variant of the logarithmic barrier method, in which we are allowed to work with a (small) subset of all the constraints. This cuts down the number of operations per iteration needed to compute the normal matrix.

4.1 Approximate solutions and rank–one updates

4.1.1 The revised logarithmic barrier algorithm

As stated in Section 2.6, the total number of arithmetic operations for the logarithmic barrier method for linear programming is $O(n^{3.5} \ln \frac{n\mu_0}{\epsilon})$ for the short- and medium–step variant, and $O(n^4 \ln \frac{n\mu_0}{\epsilon})$ for the long–step variant. For finding an exact optimal solution the complexity bounds become $O(n^{3.5}L)$ and $O(n^4 L)$, respectively.

The original $O(n^{3.5}L)$ complexity bound for *short–step* path–following methods was reduced to $O(n^3 L)$ by Vaidya [149] for the center method, by Gonzaga [47] for the logarithmic barrier method and by Kojima, Mizuno and Yoshise [79] and Monteiro and Adler [110] for the primal–dual path–following method. This reduction was achieved by using Karmarkar's [72] partial updating scheme. Their partial updating analysis is based on steps of a fixed, short length, which fits in short–step methods in a natural way. In Mizuno and Todd [106] a partial updating analysis for an 'adaptive step' path–following algorithm is given.

111

As mentioned above, the partial updating analysis in [149], [47], [79] and [110] is based on steps of a short, fixed length, and so it cannot be used in medium– or long–step algorithms. In this section we show that by using a Goldstein–Armijo rule to safeguard the line searches of the barrier function, a $\sqrt{n}$ reduction in the complexity bounds can be obtained for both variants. The Goldstein–Armijo rule was introduced in the complexity analysis for Karmarkar's [72] projective algorithm by Anstreicher [3]. Anstreicher and Bosch [6] used the rule to improve the complexity bound for the affine potential reduction algorithms in Ye [162] and Freund [36].

We consider the dual linear programming problem $(\mathcal{LD})$ as given in Section 2.3. We will also use some of the results obtained in that section. In each iteration we have to solve p from

$$Hp = -g.$$

So, essentially, in each iteration of the logarithmic barrier method, the $m \times m$ coefficient matrix $H = AS^{-2}A^T$ of this system has to be inverted (see (2.14)). Hence, assuming that $m = O(n)$, at each iteration $O(n^3)$ arithmetic operations are needed. The matrices in two successive iterations differ only due to changes in S. Now consider the hypothetical case when only one entry of S changes. Then the new coefficient matrix H' differs from H only by a rank–one matrix. So we can write

$$H' = H + uv^T,$$

for suitable vectors u and v. With the help of the Sherman–Morrison formula (see e.g. Golub and van Loan [46]) we have for the inverse of H'

$$(H + uv^T)^{-1} = H^{-1} - \frac{H^{-1}uv^T H^{-1}}{1 + v^T H^{-1} u}. \tag{4.1}$$

Since in (4.1) only matrix–vector multiplications are involved, the inverse of H' can be calculated from the inverse of H in only $O(n^2)$ arithmetic operations. If we require an exact solution of the system of equations we will in general need to make n such rank–one modifications, since all n components of s may change. Therefore $O(n^3)$ arithmetic operations will be needed at each iteration.

However, assume that instead of $AS^{-2}A^T$ we use $A\tilde{S}^{-2}A^T$, where $\tilde{S}$ is a working matrix closely related to S. Actually, the diagonal term $\tilde{s}_j$ of $\tilde{S}$ is updated during the inner iteration only if $\tilde{s}_j$ differs too much from s_j. If a limited number of components of $\tilde{s}$ are updated at a given iteration, a reduced computational cost can be achieved using the Sherman–Morrison formula. Of course one does not obtain the exact projected Newton direction p, but an approximation $\tilde{p}$ of it. More precisely, an explicit expression for $\tilde{p}$ is (cf. (2.14))

$$\tilde{p} = (A\tilde{S}^{-2}A^T)^{-1}\left(\frac{b}{\mu} - AS^{-1}e\right). \tag{4.2}$$

The purpose is now to show that by performing a safeguarded line search along $\tilde{p}$, one can achieve the double goal of enforcing a significant decrease of the barrier function at each iteration, while maintaining the number of updates in the components of $\tilde{s}$ relatively small, thereby achieving a computational saving.

In order to work out these ideas we introduce the diagonal matrix D, with diagonal element d_j, defined by

$$\tilde{S} = SD. \tag{4.3}$$

Let $\rho > 1$ be some fixed number. The algorithm is designed so as to maintain the inequality

$$\frac{1}{\rho} \leq d_i \leq \rho, \ 1 \leq i \leq n. \tag{4.4}$$

Karmarkar [72] already used approximate solutions and partial updating to reduce the complexity bound for his algorithm. Using these approximate solutions for S, we will show that on the average only $\sqrt{n}$ rank–one modifications are needed, without increasing the complexity bound for the required number of iterations. This can be reached by submitting the line search to a Goldstein–Armijo condition.

To measure the distance to the central path, we shall use a slightly different version of the δ–metric (2.12), introduced in Section 2.3. We define

$$\tilde{\delta}(y,\mu) := \min_{x} \left\{ \|D(\frac{Sx}{\mu} - e)\| : Ax = b \right\}. \tag{4.5}$$

Again, there is a close relationship between this measure and the approximate Newton direction $\tilde{p}$. It can easily be verified that (cf. Lemma 2.2)

$$\tilde{\delta}(y,\mu) = \|\tilde{S}^{-1}A^T\tilde{p}\|. \tag{4.6}$$

It is clear from the definition that $\tilde{\delta}(y,\mu) = 0$ if and only if $y = y(\mu)$. In other words, we will have

$$\delta(y,\mu) = 0 \iff \tilde{\delta}(y,\mu) = 0.$$

It is easy to verify that under assumption (4.4)

$$\frac{1}{\rho}\delta(y,\mu) \leq \tilde{\delta}(y,\mu) \leq \rho\delta(y,\mu). \tag{4.7}$$

Consequently, if $\tilde{\delta}(y,\mu) \leq \frac{1}{\rho}$ then we have $\delta(y,\mu) \leq 1$, and the lemmas proved in the Section 2.3 hold.

The Goldstein–Armijo condition can be formulated as follows:

$$\frac{\Delta\phi_\text{B}}{\alpha} \geq -\zeta \left. \frac{d\phi_\text{B}(y + \alpha\tilde{p},\mu)}{d\alpha} \right|_{\alpha=0}, \tag{4.8}$$

where $\Delta\phi_\text{B}$ is the change in the barrier function value and $0 < \zeta < 1$. This condition is a well-known rule in nonlinear programming; see e.g. [101]. It says that the step length must be such that the decrease in the logarithmic barrier function value divided by the step length is greater than a constant times the derivative at y. So it permits significant decreases of $\phi_B(y, \mu)$, but prevents excessively long steps. Note that there are always values for α which satisfy condition (4.8), since

$$\lim_{\alpha \to 0} \frac{\Delta \phi_B}{\alpha} = -\left.\frac{d\phi_B(y + \alpha \tilde{p}, \mu)}{d\alpha}\right|_{\alpha=0},$$

and $0 < \zeta < 1$. Also note that we have

$$
\begin{aligned}
\left.\frac{d\phi_B(y + \alpha \tilde{p}, \mu)}{d\alpha}\right|_{\alpha=0} &= g^T \tilde{p} \\
&= -(\frac{b}{\mu} - AS^{-1}e)^T \tilde{p} \\
&= -\|\tilde{S}^{-1} A^T \tilde{p}\|^2 \\
&= -\tilde{\delta}(y, \mu)^2, \quad\quad\quad (4.9)
\end{aligned}
$$

where the second equality follows from (4.2) and the last equality from (4.6).

We will now describe the revised algorithm to find an ϵ-optimal solution.

Revised Logarithmic Barrier Algorithm

Input:
ϵ is the accuracy parameter;
θ is the reduction parameter, $0 < \theta < 1$;
ρ is the coordinate update parameter, $\rho > 1$;
ζ is the Goldstein–Armijo factor, $\zeta \leq \frac{1}{2}$;
μ_0 is the initial barrier value;
y^0 is a given interior feasible point such that $\tilde{\delta}(y^0, \mu_0) \leq \frac{1}{2\rho}$;

begin
$\quad y := y^0; \ \tilde{s} := s^0; \ \mu := \mu_0;$
$\quad$ **while** $\mu > \frac{\epsilon}{4n}$ **do**
$\quad$ **begin** (outer step)
$\quad\quad \mu := (1 - \theta)\mu;$
$\quad\quad$ **while** $\tilde{\delta}(y, \mu) > \frac{1}{2\rho}$ **do**
$\quad\quad$ **begin** (inner step)
$\quad\quad\quad D := \tilde{S}S^{-1}$
$\quad\quad\quad \tilde{\alpha} := \arg\min_{\alpha>0}\left\{\phi_B(y + \alpha \tilde{p}, \mu) : s - \alpha A^T \tilde{p} > 0, \ \frac{\Delta \phi_B}{\alpha} \geq \zeta \tilde{\delta}(y, \mu)^2\right\}$
$\quad\quad\quad y := y + \tilde{\alpha}\tilde{p}$
$\quad\quad\quad$ **for** $j := 1$ **to** n **do if** $\frac{\tilde{s}_j}{s_j} \notin (\frac{1}{\rho}, \rho)$ **then** $\tilde{s}_j := s_j$
$\quad\quad$ **end** (inner step)
$\quad$ **end** (outer step)
end.

The line search need not be exact; a suitable steplength will be sufficient for the complexity analysis.

4.1.2   Complexity analysis

We first give upper bounds for the total number of outer and inner iterations. Finally, we derive an upper bound for the total number of coordinate updates of $\tilde{s}$.

Henceforth we shall denote $\{y^j\}$, $j = 0, 1, 2, \cdots$ the sequence of inner iterates and $\{\mu_k\}$, $k = 0, 1, 2, \cdots$ the sequence of parameter values during the successive outer iterations. Suppose that y^j is the current iterate when μ_k is calculated. Then set $m_k = j$. Take $m_0 = 0$. Then for any $j > 0$ there is a k such that $m_k < j \leq m_{k+1}$, and the value of μ used in the calculation of y^j is $\mu_k = (1 - \theta)^k \mu_0$.

Theorem 4.1 *After at most*

$$\frac{1}{\theta} \ln \frac{4n\mu_0}{\epsilon}$$

outer iterations, the Revised Logarithmic Barrier Algorithm ends up with a solution such that $z^* - b^T y \leq \epsilon$.

Proof: Since $\tilde{\delta}(y, \mu) \leq \frac{1}{2\rho}$ implies $\delta(y, \mu) \leq \frac{1}{2}$, the theorem is an immediate consequence of Theorem 2.2.   □

Lemma 4.1 *Let* $\tilde{\delta} := \tilde{\delta}(y, \mu)$, $\bar{\alpha} := [\rho(\tilde{\delta} + \rho)]^{-1}$. *Then*

$$\bar{\Delta}\phi_{\mathrm{B}} := \phi_{\mathrm{B}}(y, \mu) - \phi_{\mathrm{B}}(y + \bar{\alpha}\tilde{p}, \mu) \geq \frac{\tilde{\delta}}{\rho} - \ln(1 + \frac{\tilde{\delta}}{\rho}).$$

Moreover $\frac{\bar{\Delta}\phi_{\mathrm{B}}}{\bar{\alpha}} \geq \zeta\tilde{\delta}^2$, *for* $\zeta \leq \frac{1}{2}$.

Proof: The first part can easily be proved by modifying the proof of Lemma 2.6 appropriately: We write down the Taylor series for ϕ_{B} with respect to α:

$$\phi_{\mathrm{B}}(y + \alpha\tilde{p}, \mu) = \phi_{\mathrm{B}}(y, \mu) + \alpha g^T \tilde{p} + \frac{1}{2}\alpha^2 \tilde{p}^T H \tilde{p} + \sum_{k=3}^{\infty} t_k,$$

where t_k denotes the k-th order term in the Taylor series. We will also use the notation a_i for the i-th column of A. Since the k-th order differential of $-\ln s_i$ at y and $\tilde{p}$ is equal to

$$(-1)^k \left(\frac{a_i^T \tilde{p}}{s_i}\right)^k (k - 1)!,$$

we have

$$
\begin{aligned}
t_k &= \frac{\alpha^k}{k!} \sum_{i=1}^{n} (-1)^k \left(\frac{a_i^T \tilde{p}}{s_i}\right)^k (k - 1)! \\
&= \frac{(-\alpha)^k}{k} \sum_{i=1}^{n} \left(\frac{a_i^T \tilde{p}}{s_i}\right)^k.
\end{aligned}
$$

Hence, we have

$$
\begin{aligned}
|t_k| &\leq \frac{\alpha^k}{k} \sum_{i=1}^{n} \left| \frac{a_i^T \tilde{p}}{s_i} \right|^k \\
&\leq \frac{\alpha^k}{k} \left(\sum_{i=1}^{n} \left| \frac{a_i^T \tilde{p}}{s_i} \right|^2 \right)^{\frac{k}{2}} \\
&= \frac{\alpha^k}{k} \left(\sum_{i=1}^{n} d_i^2 \left| \frac{a_i^T \tilde{p}}{\tilde{s}_i} \right|^2 \right)^{\frac{k}{2}} \\
&\leq \frac{\alpha^k}{k} \left(\sum_{i=1}^{n} \rho^2 \left| \frac{a_i^T \tilde{p}}{\tilde{s}_i} \right|^2 \right)^{\frac{k}{2}} \\
&= \frac{\alpha^k}{k} (\rho \tilde{\delta})^k,
\end{aligned}
$$

where the first equality follows from (4.3), the last inequality from (4.4) and the last equality from (4.6). So, since

$$
g^T \tilde{p} = -\tilde{\delta}^2
$$

(equation (4.9), we find

$$
\begin{aligned}
\phi_{\text{B}}(y + \alpha \tilde{p}, \mu) &\leq \phi_{\text{B}}(y, \mu) - \alpha \tilde{\delta}^2 + \sum_{k=2}^{\infty} \frac{\alpha^k}{k} (\rho \tilde{\delta})^k \\
&= \phi_{\text{B}}(y, \mu) - \alpha \tilde{\delta}^2 - \ln(1 - \alpha \rho \tilde{\delta}) - \alpha \rho \tilde{\delta}.
\end{aligned}
$$

The last equality only holds if

$$
\alpha \rho \tilde{\delta} < 1. \tag{4.10}
$$

Hence

$$
\bar{\Delta} \phi_{\text{B}} \geq \alpha(\tilde{\delta}^2 + \rho \tilde{\delta}) + \ln(1 - \alpha \rho \tilde{\delta}). \tag{4.11}
$$

The right–hand side is maximal if

$$
\alpha = \bar{\alpha} = \frac{1}{\rho(\tilde{\delta} + \rho)}.
$$

This value for α also satisfies condition (4.10). Substitution of this value finally gives

$$
\bar{\Delta} \phi_{\text{B}} \geq \frac{\tilde{\delta}}{\rho} - \ln(1 + \frac{\tilde{\delta}}{\rho}).
$$

This proves the first part of the lemma.

The second part follows immediately from Lemma B.5:

$$
\bar{\Delta} \phi_{\text{B}} \geq \frac{\tilde{\delta}}{\rho} - \ln(1 + \frac{\tilde{\delta}}{\rho})
$$

$$
\geq \frac{\tilde{\delta}}{\rho} - \frac{\tilde{\delta}}{\rho} + \frac{\frac{\tilde{\delta}^2}{\rho^2}}{2(1 + \frac{\tilde{\delta}}{\rho})}
$$

$$
= \frac{\tilde{\delta}^2}{2\rho(\tilde{\delta} + \rho)} \tag{4.12}
$$

$$
= \frac{1}{2}\bar{\alpha}\tilde{\delta}^2.
$$

$\square$

Theorem 4.2 *Each outer iteration of the revised algorithm requires at most*

$$
\frac{12\theta\rho^4}{(1 - \theta)^2}\left(\theta n + \frac{3}{2}\sqrt{n}\right) + 4\rho^4
$$

inner iterations.

Proof: Let us consider the $(k + 1)$'st outer iteration. Let N denote the number of inner iterations. For each inner iteration we know, by the definition of $\tilde{\alpha}$ and (4.12), that the decrease in the barrier function value is larger than $\frac{\tilde{\delta}^2}{2\rho(\tilde{\delta}+\rho)}$. Since this expression is an increasing function of $\tilde{\delta}$, and since during each iteration $\tilde{\delta} \geq \frac{1}{2\rho}$, we have

$$
\frac{\tilde{\delta}^2}{2\rho(\tilde{\delta} + \rho)} \geq \frac{1}{12\rho^4}.
$$

The difference of the logarithmic barrier function value in the current iterate y^{m_k} and the next center $y(\mu_k)$ is equal to

$$
\phi_B(y^{m_k}, \mu_k) - \phi_B(y(\mu_k), \mu_k).
$$

Consequently, we have

$$
\frac{1}{12\rho^4}N \leq \phi_B(y^{m_k}, \mu_k) - \phi_B(y(\mu_k), \mu_k). \tag{4.13}
$$

From the proof of Theorem 2.3 it follows that

$$
\phi_B(y^{m_k}, \mu_k) - \phi_B(y(\mu_k), \mu_k) \leq \frac{1}{3} + \frac{\theta}{(1 - \theta)^2}\left(\theta n + \frac{3}{2}\sqrt{n}\right). \tag{4.14}
$$

The theorem follows by combining (4.13) and (4.14). $\square$

As a consequence of Lemma 4.1 and Theorem 4.2 we have that using an additional Goldstein–Armijo rule and approximate solutions do not influence the order of the total number of outer and inner iterations.

The last theorem will give an upper bound for the total number of coordinate updates in $\tilde{s}$. For the proof of this theorem we make use of some results obtained by Anstreicher [3].

Theorem 4.3 *The total number W of coordinate updates of $\tilde{s}$ up to the last inner iteration is bounded by*

$$W \leq \frac{2\rho^3 \sqrt{n}}{\zeta(\rho - 1)} \left(\frac{1}{3\theta} + \frac{1}{(1-\theta)^2} \left(\theta n + \frac{3}{2} \sqrt{n} \right) \right) \ln \frac{4n\mu_0}{\epsilon}.$$

Proof: Let[1] K denote the number of outer iterations, and M the total number of inner iterations. Let k_1 be an iteration at which an update of $\tilde{s}_i$ is performed. Let $k_2 > k_1$ be the first iteration at which $\tilde{s}_i$ is updated again. Then we have

$$\prod_{j=k_1+1}^{k_2} \max\left(\frac{s_i^j}{s_i^{j-1}}, \frac{s_i^{j-1}}{s_i^j} \right) \geq \max\left(\prod_{j=k_1+1}^{k_2} \frac{s_i^j}{s_i^{j-1}}, \prod_{j=k_1+1}^{k_2} \frac{s_i^{j-1}}{s_i^j} \right)$$

$$= \max\left(\frac{s_i^{k_2}}{s_i^{k_1}}, \frac{s_i^{k_1}}{s_i^{k_2}} \right)$$

$$\geq \rho.$$

Taking logarithms and defining

$$r_i^j := 1 - \tilde{\alpha}_j (s_i^j)^{-1} a_i^T \tilde{p}^j = \frac{s_i^{j+1}}{s_i^j},$$

we obtain

$$\ln \rho \leq \sum_{j=k_1}^{k_2-1} |\ln r_i^j|. \tag{4.15}$$

Letting

$$\hat{r}_i^j = \max\{r_i^j, \frac{1}{\rho}\},$$

inequality (4.15) can be sharpened to

$$\ln \rho \leq \sum_{j=k_1}^{k_2-1} |\ln \hat{r}_i^j|. \tag{4.16}$$

To prove (4.16), first assume that for some ℓ, $k_1 \leq \ell \leq k_2 - 1$, $r_i^j < \frac{1}{\rho}$. Then

$$\ln \rho = |\ln \hat{r}_i^\ell| \leq \sum_{j=k_1}^{k_2-1} |\ln \hat{r}_i^j|.$$

Else $\hat{r}_i^j = r_i^j$, $k_1 \leq j \leq k_2 - 1$, and (4.16) holds because of (4.15). Hence, (4.16) has been proved. We deduce from (4.16) a bound on the total number w_i of updates of coordinate i of $\tilde{s}$:

$$w_i \ln \rho \leq \sum_{j=0}^{M-1} |\ln \hat{r}_i^j|.$$

---

[1] I would like to thank Kurt Anstreicher for the valuable suggestions for the proof of Theorem 4.3 he made during a visit to Delft, May 14–24, 1990.

Consequently the total number of coordinate updates is bounded by

$$W \ln \rho \leq \sum_{j=0}^{M-1} \sum_{i=1}^{n} |\ln \hat{r}_i^j|. \tag{4.17}$$

In view of Lemma B.6, with $v = \hat{r}_i^k$ and $w = \frac{1}{\rho}$,

$$|\ln \hat{r}_i^j| \leq \frac{\ln \rho}{1 - \frac{1}{\rho}} |1 - \hat{r}_i^j|. \tag{4.18}$$

Since $\hat{r}_i^j = r_i^j$ if $r_i^j \geq \frac{1}{\rho}$, and $\hat{r}_i^j > r_i^j$ if $r_i^j < \frac{1}{\rho}$ we always have

$$|1 - \hat{r}_i^j| \leq |1 - r_i^j| = \tilde{\alpha}_j |(s_i^j)^{-1} a_i^T \tilde{p}^j|. \tag{4.19}$$

Substitution of (4.18) and (4.19) into (4.17) gives

$$W \leq \frac{\rho}{\rho - 1} \sum_{j=0}^{M-1} \tilde{\alpha}_j \sum_{i=1}^{n} |(s_i^j)^{-1} a_i^T \tilde{p}^j|.$$

From the inequality relating the l_1 and l_2 norms

$$\sum_{i=1}^{n} |(s_i^j)^{-1} a_i^T \tilde{p}^j| \leq \sqrt{n} \|(S^j)^{-1} A^T \tilde{p}^j\|$$

$$\leq \rho \sqrt{n} \|(\tilde{S}^j)^{-1} A^T \tilde{p}^j\|.$$

Hence

$$W \leq \frac{\rho^2 \sqrt{n}}{\rho - 1} \sum_{j=0}^{M-1} \tilde{\alpha}_j \|(\tilde{S}^j)^{-1} A^T \tilde{p}^j\|. \tag{4.20}$$

Since the Goldstein–Armijo condition is satisfied in each inner iteration, it holds for any j and k such that $m_k < j \leq m_{k+1}$ (we will write $k(j)$ instead of k to denote its dependence on j) that

$$\tilde{\alpha}_j \leq \frac{\phi_B(y^j, \mu_{k(j)}) - \phi_B(y^{j+1}, \mu_{k(j)})}{\zeta \|(\tilde{S}^j)^{-1} A^T \tilde{p}^j\|^2}. \tag{4.21}$$

Substituting this into (4.20), we obtain

$$W \leq \frac{\rho^2 \sqrt{n}}{\zeta(\rho - 1)} \sum_{j=0}^{M-1} \frac{\phi_B(y^j, \mu_{k(j)}) - \phi_B(y^{j+1}, \mu_{k(j)})}{\|(\tilde{S}^j)^{-1} A^T \tilde{p}^j\|}.$$

Since $\|(\tilde{S}^j)^{-1} A^T \tilde{p}^j\| \geq \frac{1}{2\rho}$, this implies

$$
\begin{aligned}
W &\leq \frac{2\rho^3 \sqrt{n}}{\zeta(\rho - 1)} \sum_{j=0}^{M-1} \left(\phi_B(y^j, \mu_{k(j)}) - \phi_B(y^{j+1}, \mu_{k(j)}) \right) \\
&= \frac{2\rho^3 \sqrt{n}}{\zeta(\rho - 1)} \sum_{k=0}^{K-1} \left(\phi_B(y^{m_k}, \mu_k) - \phi_B(y^{m_{k+1}}, \mu_k) \right) \\
&\leq \frac{2\rho^3 \sqrt{n}}{\zeta(\rho - 1)} K \left(\frac{1}{3} + \frac{\theta}{(1 - \theta)^2} \left(\theta n + \frac{3}{2} \sqrt{n} \right) \right) \\
&\leq \frac{2\rho^3 \sqrt{n}}{\zeta(\rho - 1)} \left(\frac{1}{3\theta} + \frac{1}{(1 - \theta)^2} \left(\theta n + \frac{3}{2} \sqrt{n} \right) \right) \ln \frac{4n\mu_0}{\epsilon},
\end{aligned}
$$

where the last two inequalities follow from (4.14) and Theorem 4.1.   □

Theorem 4.1 and Theorem 4.2 imply that M, the total number of inner iterations needed by the algorithm, is at most

$$\left(\frac{12\rho^4}{(1-\theta)^2}\left(\theta n + \frac{3}{2}\sqrt{n}\right) + \frac{4\rho^4}{\theta}\right)\ln\frac{4n\mu_0}{\epsilon}.$$

The total number of arithmetic operations in each iteration, aside from the work due to coordinate updates, is $O(n^2)$. The same amount of work is required for each coordinate update. Consequently, the total number of arithmetic operations needed by the algorithm is $(M + W)O(n^2)$, where W is the total number of coordinate updates given by Theorem 4.3.

We conclude that to obtain an ϵ–optimal solution the Revised Logarithmic Barrier Algorithm needs

- $O(n\ln\frac{n\mu_0}{\epsilon})$ Newton iterations and $O(n^{3.5}\ln\frac{n\mu_0}{\epsilon})$ arithmetic operations for the long–step variant;

- $O(\sqrt{n}\ln\frac{n\mu_0}{\epsilon})$ Newton iterations and $O(n^3\ln\frac{n\mu_0}{\epsilon})$ arithmetic operations for the medium–step variant.

To obtain an optimal solution we have to take $\epsilon = 2^{-2L}$. For this value of ϵ the revised algorithm needs $O(n^{3.5}L)$ and $O(n^3L)$ arithmetic operations respectively, assuming that $\mu_0 \leq 2^{O(L)}$.

4.1.3   An illustrative example

In this subsection we will illustrate the Revised Logarithmic Barrier Algorithm by means of a concrete example. Consider the following simple linear programming problem, which is the same problem as considered in Section 2.3.3:

$$\begin{cases} \max\ y_1 + 2y_2 \\ y_1 \leq 0 \\ y_2 \leq 0. \end{cases}$$

In this example we will use $\theta = \frac{3}{4}$ and $\rho = 2$, and as starting point we will use $\mu^0 = 4$ and $y^0 = (-4, -\frac{5}{2})$. Note that $\tilde{s}^0 = -y^0$. In the algorithm we will use the step length given in Lemma 4.1

$$\alpha = \frac{1}{\rho(\tilde{\delta} + \rho)}.$$

According to (4.2) we have that

$$\begin{aligned}
\tilde{p}(y,\mu) &= (A\tilde{S}^{-2}A^T)^{-1}\left(\frac{b}{\mu} - AS^{-1}e\right) \\
&= \begin{pmatrix} (\tilde{s}_1)^2(\frac{1}{\mu} + \frac{1}{y_1}) \\ (\tilde{s}_2)^2(\frac{2}{\mu} + \frac{1}{y_2}) \end{pmatrix}.
\end{aligned} \tag{4.22}$$

Moreover

$$
\begin{aligned}
\tilde{\delta}(y,\mu) &= \left\| \tilde{S}^{-1} A^T \tilde{p}(y,\mu) \right\| \\
&= \left\| \begin{array}{c} \tilde{s}_1(\frac{1}{\mu} + \frac{1}{y_1}) \\ \tilde{s}_2(\frac{2}{\mu} + \frac{1}{y_2}) \end{array} \right\|.
\end{aligned}
\tag{4.23}
$$

It is easy to verify that $\tilde{\delta}(y^0, \mu^0) = \frac{1}{4}$. Hence in the algorithm we can reduce μ:

$$
\mu^1 = (1 - \frac{3}{4})\mu^0 = 1.
$$

Hence we have

$$
\tilde{p}(y^0, \mu^1) = \begin{pmatrix} 12 \\ 10 \end{pmatrix},
$$

and $\tilde{\delta}(y^0, \mu^1) = 5$. For the step length we have

$$
\alpha = \frac{1}{\rho(\tilde{\delta} + \rho)} = \frac{1}{14}.
$$

So, the next iterate is

$$
\begin{aligned}
y^1 &= y^0 + \alpha\tilde{p}(y^0, \mu^1) \\
&\approx -\begin{pmatrix} 3.14 \\ 1.79 \end{pmatrix}.
\end{aligned}
$$

Since $\frac{4}{3.14}$ and $\frac{2.5}{1.79}$ are larger than $\frac{1}{2}$ and less 2, we have that $\tilde{s}^1 = \tilde{s}^0$. Using (4.23) we obtain $\tilde{\delta}(y^1, \mu^0) \approx 4.51$. Hence we take another Newton step. Again using (4.22) we have

$$
\tilde{p}(y^1, \mu^1) \approx \begin{pmatrix} 10.91 \\ 9 \end{pmatrix},
$$

and $\alpha \approx \frac{1}{2(4.51+2)} \approx 0.0767$. So, we obtain

$$
y^2 \approx -\begin{pmatrix} 2.31 \\ 1.10 \end{pmatrix}.
$$

For the next iterate we have to change $\tilde{s}$, since the second coordinate becomes to small ($\frac{2.5}{1.10} > 2$):

$$
\tilde{s}^2 \approx \begin{pmatrix} 4 \\ 1.10 \end{pmatrix}.
$$

Since

$$\tilde{\delta}(y^2, \mu^1) \approx 2.56 > \frac{1}{4}$$

we have to take more Newton steps. We will only give the results for the next step, and leave it to the reader to verify them:

$$\tilde{p}(y^2, \mu^1) \approx \begin{pmatrix} 9.06 \\ 1.30 \end{pmatrix},$$

and $\alpha = 0.110$, and

$$y^3 \approx - \begin{pmatrix} 1.31 \\ 0.95 \end{pmatrix}.$$

Hence

$$\tilde{s}^3 \approx \begin{pmatrix} 1.31 \\ 1.10 \end{pmatrix}.$$

4.2 Adding and deleting constraints

4.2.1 General remarks

One drawback to all interior point methods is the great computational effort required in each iteration. In each iteration the search direction p is obtained by solving a linear system with normal matrix $AD^{-2}A^T$, where A is the $(m \times n)$ constraint matrix and D a positive diagonal matrix depending on the current iterate. Therefore, working with a subset of the dual constraints rather than the full system, would save a great deal of computation, especially if $n \gg m$.

The first such an attempt to save computations is the so–called 'build–down' or 'column deletion' method, proposed by Ye [159] and [161]. In his approach, a criterion for detecting (non)binding constraints is derived on the basis of an ellipsoid which circumscribes the optimal facet. If a constraint is detected to be nonbinding in the optimal set, it is removed from the system. Consequently, the system becomes increasingly smaller, which reduces the computational effort for computing the normal matrix $AD^{-2}A^T$. However, the speed of the detection process is crucial. If the nonbinding constraints are only detected during the last stage of the algorithm, the reduction in computation is negligible. To the best of our knowledge, there are no computational results to be found in the literature concerning this build–down process.

The second attempt to save computations is the 'build–up' or 'column generation' method. Papers on column generation techniques within interior point methods were first written by Mitchell [102] and Goffin and Vial [42] for the projective method. In [8] computational results for Goffin and Vial's method applied to convex programming are reported. However, these papers provide no theoretical analysis for the effect on the potential function and/or the number of iterations after the addition of a column/row.

Ye [163] proposed a (non–interior) potential reduction method for the linear feasibility problem which allows column generation. In each iteration an inequality violated at the current center is added to the system (in a shifted position), until a feasible point has been found. He proved that such a point can be found in $O(\sqrt{n}L)$ iterations, where L is the input length of the problem. Although each linear programming problem can be formulated as a linear feasibility problem, this is an inefficient way of solving linear programming problems.

Dantzig and Ye [17] proposed a build–up scheme for the dual affine scaling algorithm. This method differs from the 'standard' affine scaling method in that the ellipsoid chosen to generate the search direction p is constructed from a set of m 'promising' dual constraints. If the next iterate $y + p$ violates one of the other constraints, this constraint is added to the current system and a new ellipsoid and search direction (using the new set of constraints) are calculated. After making the step, a new set of m promising dual constraints is selected.

Tone [142] proposed an active–set strategy for Ye's [162] dual potential reduction method. In this strategy the search direction is also constructed from a subset of constraints which have small dual slacks in the current iterate. More constraints are added if no sufficient potential reduction is obtained. After making the step a new set of dual constraints, with small slack values, is selected. This algorithm converges to an optimal solution in $O(\sqrt{n}L)$ iterations. In [70] some computational results for this active–set strategy are reported.

Elaborating the above ideas, in this section we will propose a build–up and down strategy for the logarithmic barrier method for linear programming (see Section 2.3). This strategy starts with a (small) subset of the constraints, and follows the corresponding central path until the iterate is 'close' to (or violates) one of the other constraints, at which point the constraint is added to the current system. Moreover, a constraint is deleted if the corresponding slack value of the current approximately centered iterate is 'large'. This process is repeated until the iterate is close to the optimum. We will derive an upper bound for the total number of iterations required by this build–up and down algorithm. This upper bound will appear to depend on q^*, which is the maximal number of constraints in the subsystem, rather then on n.

Comparing with the usual logarithmic barrier method, this build–up and down method has two advantages. Firstly, it has the before mentioned property of using only subsets of the set of all constraints and hence decreases the computational effort per iteration. Secondly, since it is likely that $q^* < n$, the (theoretical) complexity is better than that of the standard path–following method.

In the next subsection we first analyze the effect of shifting, adding and deleting a constraint on the position of the center, the 'distance' to the path, and the change in the barrier function. These basic properties are interesting in itself, but also enable us to analyze the proposed build–up and down algorithm.

4.2.2   The effects of shifting, adding and deleting constraints

Again, we consider the dual linear programming problem $(\mathcal{LD})$ as given in Section 2.3. We will also use some of the results obtained in that section.

In the algorithm we will add and delete constraints. In the analysis we also need to consider the effect of shifting a constraint. That is why we start with some lemmas dealing with shifting a constraint. Note that if we take $b = 0$ then some of Ye's [163] results follow from some lemmas in this section.

In the sequel of this section we will denote by $(AS^{-2}A^T)_Q$ the matrix $AS^{-2}A^T$ restricted to the columns of A in the index set Q, i.e.

$$(AS^{-2}A^T)_Q = \sum_{i \in Q} \frac{a_i a_i^T}{s_i^2}.$$

Moreover, we define

$$\|z\|_Q := \sqrt{z^T(AS^{-2}A^T)_Q^{-1}z},$$

assuming that $(AS^{-2}A^T)_Q$ has full rank. The full index set $\{1, \cdots, n\}$ is denoted by N, and

$$\delta_i := \frac{s_i}{\|a_i\|_N} = \frac{s_i}{\sqrt{a_i^T(AS^{-2}A^T)^{-1}a_i}},$$

$i = 1, \cdots, n$. Note that δ_i can be interpreted as the distance to the i-th constraint measured in a certain metric.

Shifting a constraint

Figure 4.1 shows for a special problem how the central path changes if a constraint is shifted. Suppose the first constraint is shifted by a fraction ε of the current slack s_1. So we replace the constraint

$$a_1^T y \leq c_1$$

by

$$a_1^T y \leq c_1 - \varepsilon s_1, \ 0 \leq \varepsilon \leq 1.$$

Let the asterisk $*$ refer to this new situation; so $s_1^* = (1 - \varepsilon)s_1$ and $s_i^* = s_i$ for $i = 2, \cdots, n$. The following lemma shows the effect on the δ-measure (see (2.12)).

Lemma 4.2 *Suppose that the first constraint is shifted by εs_i, a fraction ε of the current slack value. Let the asterisk $*$ refer to this situation. Then we have*

$$\delta^*(y, \mu) \leq \delta(y, \mu) + \varepsilon(1 + \delta(y, \mu)).$$

Proof: By definition (2.12)

$$\delta^*(y, \mu) = \min_{x^*}\left\{\left\|\frac{S^*x^*}{\mu} - e\right\| : Ax^* = b\right\}$$

$$= \min_{x^*}\left\{\left\|\frac{Sx^* - \varepsilon s_1 x_1^* e_1}{\mu} - e\right\| : Ax^* = b\right\}.$$

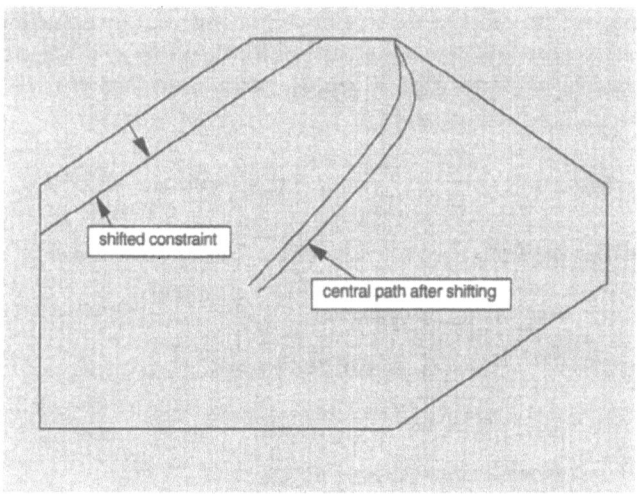

Figure 4.1: The central path before and after the constraint is shifted.

For simplicity we denote $x = x(y, \mu)$, where $x(y, \mu)$ solves the minimization problem in (2.12). Taking $x^* = x$, it follows that

$$
\begin{aligned}
\delta^*(y, \mu) &\leq \left\| \frac{Sx}{\mu} - e - \frac{\varepsilon s_1 x_1 e_1}{\mu} \right\| \\
&\leq \left\| \frac{Sx}{\mu} - e \right\| + \left\| \frac{\varepsilon s_1 x_1 e_1}{\mu} \right\| \\
&= \delta(y, \mu) + \frac{\varepsilon s_1 x_1}{\mu} \|e_1\| \\
&= \delta(y, \mu) + \frac{\varepsilon s_1 x_1}{\mu},
\end{aligned}
$$

where the first equality follows by definition (2.12). Now using that $\left| \frac{x_1 s_1}{\mu} - 1 \right| \leq \delta(y, \mu)$, and hence $\frac{s_1 x_1}{\mu} \leq 1 + \delta(y, \mu)$, we obtain

$$
\delta^*(y, \mu) \leq \delta(y, \mu) + \varepsilon(1 + \delta(y, \mu)),
$$

which proves the lemma.   □

The following lemma shows that after shifting the first constraint by an amount $\varepsilon s_1(\mu)$, the slack variable $s_1^*(\mu)$ in the new center $y^*(\mu)$ is smaller than $s_1(\mu)$, but the difference is at most the amount of shifting.

Lemma 4.3 *Suppose that the first constraint is shifted by $\varepsilon s_1(\mu)$, a fraction ε of the slack value in the μ-center. Let the asterisk $*$ refer to this situation. Then we have*

$$
(1 - \varepsilon)s_1(\mu) \leq s_1^*(\mu) \leq s_1(\mu)
$$

Proof: We analyze the effect of shifting the first constraint by an amount η. For the moment, the center for this new situation (with respect to μ) is denoted by $y(\mu, \eta)$. So $y^*(\mu) = y(\mu, \varepsilon s_1(\mu))$, and $y(\mu) = y(\mu, 0)$. First note that $y(\mu, \eta)$ is the unique minimum of

$$\phi_B^*(y, \mu) = -\frac{b^T y}{\mu} - \sum_{i=2}^{n} \ln(c_i - a_i^T y) - \ln(c_1 - a_1^T y - \eta),$$

and hence satisfies the following Karush–Kuhn–Tucker conditions:

$$\begin{cases} Ax = b \\ A^T y + \eta e_1 + s = c \\ Sx = \mu e. \end{cases} \tag{4.24}$$

Taking the derivative with respect to η, we get

$$\begin{cases} Ax' = 0 \\ A^T y' + e_1 + s' = 0 \\ Xs' + Sx' = 0, \end{cases} \tag{4.25}$$

where the prime denotes the derivative with respect to η. From (4.24) and (4.25) we derive that

$$0 = AX^2 s' = -AX^2 A^T y' - AX^2 e_1,$$

and hence $y' = -(AX^2 A^T)^{-1} AX^2 e_1$. This means that

$$s' = A^T (AX^2 A^T)^{-1} AX^2 e_1 - e_1$$

and

$$Xs' = -P_{AX}(Xe_1) = -x_1 P_{AX}(e_1).$$

Consequently,

$$s_1' = -e_1^T P_{AX}(e_1).$$

Since the eigenvalues of the projection matrix P_{AX} are zero or one, the result is that $-1 \leq s_1' \leq 0$, from which the lemma follows. $\sqcap$

The following lemma shows that the barrier function value in the μ–center increases by at least ε after shifting a constraint by a fraction ε of the corresponding slack value.

Lemma 4.4 *Suppose that the first constraint is shifted by $\varepsilon s_1(\mu)$, a fraction ε of the slack value in the μ–center. Let the asterisk $*$ refer to this situation. Then we have*

$$\phi_B^*(y^*(\mu), \mu) \geq \phi_B(y(\mu), \mu) + \varepsilon$$

Proof: Using the notation introduced in the proof of Lemma 4.3, we have

$$-\frac{d\,\phi_{\mathrm{B}}^{*}(y(\mu,\eta),\mu)}{d\,\eta} = \frac{b^T y'}{\mu} + \sum_{i=1}^{n} \frac{s_i'}{s_i}.$$

Since, by using (4.24) and (4.25),

$$\begin{aligned}
b^T y' &= x^T A^T y' \\
&= \mu e^T S^{-1} A^T y' \\
&= -\mu e^T S^{-1}(s' + e_1),
\end{aligned}$$

this results into

$$\begin{aligned}
\frac{d\,\phi_{\mathrm{B}}^{*}(y(\mu,\eta),\mu)}{d\,\eta} &= e^T S^{-1}(s' + e_1) - \sum_{i=1}^{n} \frac{s_i'}{s_i} \\
&= e^T S^{-1} e_1 \\
&= \frac{1}{s_1(\mu,\eta)}. \tag{4.26}
\end{aligned}$$

Now using the Mean Value Theorem we obtain

$$\begin{aligned}
\phi_{\mathrm{B}}^{*}(y^{*}(\mu),\mu) &= \phi_{\mathrm{B}}^{*}(y(\mu,\varepsilon s_1(\mu)),\mu) \\
&= \phi_{\mathrm{B}}^{*}(y(\mu,0),\mu) + \varepsilon s_1(\mu)\frac{d\,\phi_{\mathrm{B}}^{*}(y(\mu,\eta),\mu)}{d\,\eta}\bigg|_{0<\eta<\varepsilon s_1(\mu)} \\
&= \phi_{\mathrm{B}}(y(\mu),\mu) + \varepsilon s_1(\mu)\frac{1}{s_1(\mu,\eta)}\bigg|_{0<\eta<\varepsilon s_1(\mu)} \\
&\geq \phi_{\mathrm{B}}(y(\mu),\mu) + \varepsilon,
\end{aligned}$$

where the last equality follows from (4.26) and the last inequality follows since $s_1(\mu,\eta) \leq s_1(\mu,0) = s_1(\mu)$, for $\eta \geq 0$, according to Lemma 4.3. This proves the lemma. □

Adding a constraint

Figure 4.2 shows for a special problem how the central path changes if a constraint is added. Suppose we add the constraint $a_0^T y \leq c_0$. Let $s_0 > 0$ be the corresponding slack variable. The next lemma states what the effect is on δ.

Lemma 4.5 *Let $\delta := \delta(y,\mu)$. Let the asterisk $*$ refer to the situation after adding the constraint $a_0^T \leq c_0$. Let $s_0 > 0$ be the current slack value. Then*

$$\delta^{*}(y,\mu) \leq \begin{cases} \dfrac{1 + \delta\delta_0}{\sqrt{1+\delta_0^2}} & \text{if } \delta_0 \geq \delta \\[2mm] \sqrt{1+\delta^2} & \text{if } \delta_0 < \delta. \end{cases}$$

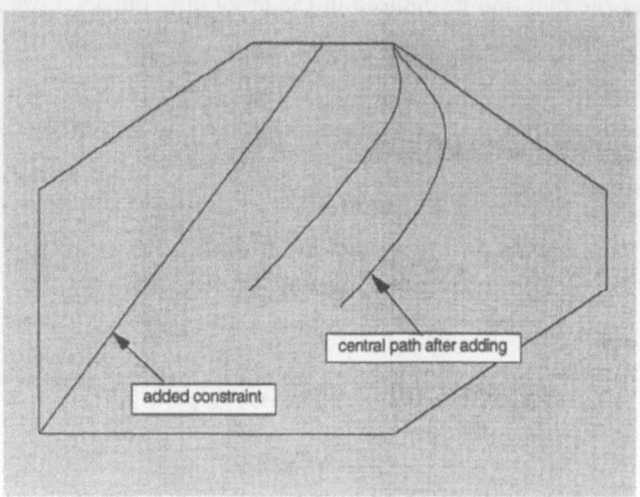

Figure 4.2: The central path before and after the constraint is added.

Proof: By definition

$$\delta^*(y,\mu)^2 = \min_{x^*,\xi}\left\{\left\|\frac{1}{\mu}\begin{pmatrix}Sx^*\\s_0\xi\end{pmatrix}-\begin{pmatrix}e\\1\end{pmatrix}\right\|^2 : Ax^*+\xi a_0 = b\right\}$$

$$= \min_{x^*,\xi}\left\{\left\|\frac{Sx^*}{\mu}-e\right\|^2+(\frac{s_0\xi}{\mu}-1)^2 : Ax^*+\xi a_0 = b\right\}.$$

Let $\Delta x := x^* - x(y,\mu)$, where $x(y,\mu)$ solves the minimization problem in (2.12). Then

$$\delta^*(y,\mu)^2 = \min_{\Delta x,\xi}\left\{\left\|\frac{Sx(y,\mu)}{\mu}-e+\frac{S\Delta x}{\mu}\right\|^2+(\frac{s_0\xi}{\mu}-1)^2 : A\Delta x = -\xi a_0\right\}$$

$$< \min_{\Delta x,\xi}\left\{\left(\left\|\frac{Sx(y,\mu)}{\mu}-e\right\|+\left\|\frac{S\Delta x}{\mu}\right\|\right)^2+(\frac{s_0\xi}{\mu}-1)^2 : A\Delta x = -\xi a_0\right\}$$

$$= \min_{\Delta x,\xi}\left\{\left(\delta+\left\|\frac{S\Delta x}{\mu}\right\|\right)^2+(\frac{s_0\xi}{\mu}-1)^2 : A\Delta x = -\xi a_0\right\}$$

$$= \min_{\Delta x,\xi}\left\{(\delta+\|S\Delta x\|)^2+(s_0\xi-1)^2 : A\Delta x = -\xi a_0\right\}$$

$$= \min_{\xi}\left\{(\delta+|\xi|\|a_0\|_N)^2+(s_0\xi-1)^2\right\},$$

where the last equality follows since

$$\Delta x = -\xi S^{-2}A^T(AS^{-2}A^T)^{-1}a_0$$

minimizes $\|S\Delta x\|^2$ over $A\Delta x = -\xi a_0$ and for this value of Δx we have

$$
\begin{aligned}
\|S\Delta x\|^2 &= |\xi| a_0^T (AS^{-2}A^T)^{-1} a_0 \\
&= |\xi| \|a_0\|_N^2.
\end{aligned}
$$

It is an easy task to prove that if $s_0 \geq \delta \|a_0\|_N$, or equivalenty $\delta_0 \geq \delta$, then the right–hand side is minimal for

$$
\xi = \frac{s_0 - \delta \|a_0\|_N}{\|a_0\|_N^2 + s_0^2} = \frac{\delta_0 - \delta}{\|a_0\|_N (1 + \delta_0^2)},
$$

else $\xi = 0$ is optimal. Substituting these values gives the lemma.   □

Note that according to Lemma 4.5, adding a constraint hardly influences the δ–measure if δ_0 is large, i.e. if the constraint which is added is far away from the current iterate. The next lemma states that if we add a constraint, then the corresponding slack value in the new center is larger than in the old center.

Lemma 4.6 *Let the asterisk $*$ refer to the situation after adding the constraint $a_0^T \leq c_0$. Then*

$$
s_0^*(\mu) \geq s_0(\mu)
$$

Proof: First note that $y^*(\mu)$ is feasible for the old situation. Moreover, if $s_0(\mu) \leq 0$, then the lemma is obviously true, since $s_0^*(\mu) > 0$. So, we may assume that $s_0(\mu) > 0$, i.e. $y(\mu)$ is also feasible for the new situation. Since $y(\mu)$ minimizes $\phi_B(y, \mu)$ we have

$$
\phi_B(y(\mu), \mu) \leq \phi_B(y^*(\mu), \mu).
$$

Since $y^*(\mu)$ minimizes $\phi_B^*(y, \mu)$ we have

$$
\phi_B^*(y^*(\mu), \mu) \leq \phi_B^*(y(\mu), \mu).
$$

Now it is clear that

$$
\begin{aligned}
\ln s_0(\mu) &= \phi_B(y(\mu), \mu) - \phi_B^*(y(\mu), \mu) \\
&\leq \phi_B(y(\mu), \mu) - \phi_B^*(y^*(\mu), \mu) \\
&\leq \phi_B(y^*(\mu), \mu) - \phi_B^*(y^*(\mu), \mu) \\
&= \ln s_0^*(\mu),
\end{aligned}
$$

which means that $s_0^*(\mu) \geq s_0(\mu)$.   □

The next lemma gives an upper bound for the barrier function value in the new center after adding the constraint.

Lemma 4.7 *Let $\delta := \delta(y, \mu) \leq \frac{1}{4}$. Let the asterisk $*$ refer to the situation after adding the constraint $a_0^T \leq c_0$. Let $s_0 > 0$ be the current slack value. Then*

$$
\phi_B^*(y, \mu) - \phi_B^*(y^*(\mu), \mu) \leq \frac{1}{3} + \max\left(0, \ln \frac{4}{\delta_0}\right).
$$

Proof: Suppose that $\delta_0 \geq 4$. Then according to Lemma 4.5

$$
\begin{aligned}
\delta^*(y,\mu) &\leq \frac{1+\delta\delta_0}{\sqrt{1+\delta_0^2}} \\
&\leq \delta + \frac{1}{\sqrt{1+\delta_0^2}} \\
&\leq \frac{1}{4} + \frac{1}{\sqrt{1+16}} \\
&< \frac{1}{2}.
\end{aligned}
$$

Consequently, we have due to Lemma 2.4

$$
\phi_{\mathrm{B}}^*(y,\mu) - \phi_{\mathrm{B}}^*(y^*(\mu),\mu) \leq \frac{1}{3}. \tag{4.27}
$$

Now suppose that $\delta_0 < 4$. Then we add the constraint in a shifted position such that $\delta_0 = \frac{s_0}{\|a_0\|_N}$ is exactly 4. Let us refer to this situation by using the superscript 0. Now we may write

$$
\begin{aligned}
\phi_{\mathrm{B}}^*(y,\mu) - \phi_{\mathrm{B}}^*(y^*(\mu),\mu) &= \phi_{\mathrm{B}}^0(y^0(\mu),\mu) - \phi_{\mathrm{B}}^*(y^*(\mu),\mu) \\
&+ \phi_{\mathrm{B}}^0(y,\mu) - \phi_{\mathrm{B}}^0(y^0(\mu),\mu) \\
&+ \phi_{\mathrm{B}}^*(y,\mu) - \phi_{\mathrm{B}}^0(y,\mu). \tag{4.28}
\end{aligned}
$$

Now we deal with the three pairs of terms in the right–hand side of (4.28) separately.

The first term $\phi_{\mathrm{B}}^0(y^0(\mu),\mu) - \phi_{\mathrm{B}}^*(y^*(\mu),\mu)$ is smaller than or equal to 0, since according to Lemma 4.4 the barrier function value in the exact center increases after shifting.

Note again that according to Lemma 4.5

$$
\delta^0(y,\mu) \leq \frac{1}{4} + \frac{1}{\sqrt{1+16}} < \frac{1}{2}.
$$

From Lemma 2.4 it now follows that

$$
\phi_{\mathrm{B}}^0(y,\mu) - \phi_{\mathrm{B}}^0(y^0(\mu),\mu) \leq \frac{1}{3}.
$$

For the third term we simply have

$$
\begin{aligned}
\phi_{\mathrm{B}}^*(y,\mu) - \phi_{\mathrm{B}}^0(y,\mu) &= \ln s_0^0 - \ln s_0 \\
&= \ln \frac{4\|a_0\|_N}{s_0} \\
&= \ln \frac{4}{\delta_0}.
\end{aligned}
$$

Substituting all this into (4.28) yields

$$\phi_B^*(y,\mu) - \phi_B^*(y^*(\mu),\mu) \le \frac{1}{3} + \ln \frac{4}{\delta_0}. \tag{4.29}$$

Combining (4.27) and (4.29) gives the lemma. □

Deleting a constraint

In this section we consider the case that the constraint $a_j^T y \le c_j$ is removed from the given problem, while assuming that the remaining constraint matrix has still full rank. As before, let the asterisk $*$ refer to the new situation. Now, letting $N^* := N \setminus \{j\}$, we have the following lemma, which will be needed in Lemma 4.9.

Lemma 4.8 *Suppose the j–th constraint with slack value s_j is deleted. Let the asterisk $*$ denote this situation and let $\delta_j := \dfrac{s_j}{\|a_j\|_N}$ and $\delta_j^* := \dfrac{s_j}{\|a_j\|_{N^*}}$. Then $\delta_j > 1$ and*

$$\delta_j^* = \sqrt{\delta_j^2 - 1}.$$

Proof: To simplify the notation we will assume that $s = e$. This amounts to rescaling the columns of A by the slack values, and gives no loss of generality. Let A^* denote the matrix obtained from A by removing the j–th column. Then

$$A^*(A^*)^T = AA^T - a_j a_j^T.$$

Since $A^*(A^*)^T$ is invertible, Sherman–Morrison's formula states that

$$(A^*(A^*)^T)^{-1} = (AA^T)^{-1} + \frac{(AA^T)^{-1} a_j a_j^T (AA^T)^{-1}}{1 - a_j^T (AA^T)^{-1} a_j}.$$

Multiplying this from the left with a_j^T and from the right by a_j we obtain

$$\begin{aligned}
\|a_j\|_{N^*}^2 &= \|a_j\|_N^2 + \frac{\|a_j\|_N^4}{1 - \|a_j\|_N^2} \\
&= \frac{\|a_j\|_N^2}{1 - \|a_j\|_N^2}.
\end{aligned}$$

This can be rewritten as

$$\delta_j^{*2} = \delta_j^2 \left(1 - \frac{1}{\delta_j^2} \right) = \delta_j^2 - 1.$$

Note that certainly $\delta_j \ge 1$. To prove that $\delta_j > 1$, note that $\delta_j = 1$ if and only if $a_j^T (AA^T)^{-1} a_j = 1$. This can be written as $e_j A^T (AA^T)^{-1} A e_j = 1$. Since $A^T (AA^T)^{-1} A e_j$ equals the orthogonal projection of e_j on the row space of A, we conclude that $\delta_j = 1$ holds if and only if e_j is in the row space of A. This is equivalent to having $x_j = 0$ whenever $Ax = 0$. In other words, no dependence relation between the columns of A contains the j-th column with nonzero coefficient. From this we conclude that $\delta_j = 1$ if and only if removing the j-th column from A decreases the rank. This completes the proof of the lemma, because we assumed that the remaining constraint matrix has still full rank. □

Lemma 4.9 *Suppose the j-th constraint with slack value s_j is deleted. Let the asterisk $*$ denote this situation and let $\delta := \delta(y, \mu)$. Then*

$$\delta^*(y, \mu) \leq \delta + \frac{1 + \delta}{\delta_j^*} = \delta + \frac{1 + \delta}{\sqrt{\delta_j^2 - 1}}.$$

Proof: Again let A^* (S^*) denote the matrix obtained from A (S) by removing the j-th column. Then by definition

$$\delta^*(y, \mu) = \min_{x^*} \left\{ \left\| \frac{S^* x^*}{\mu} - e \right\| : A^* x^* = b \right\}. \tag{4.30}$$

Let $x(y, \mu)^T = (\xi, x^*(y, \mu)^T)$ solve the minimization problem for $\delta(y, \mu)$. Then

$$\begin{aligned}
\delta^2 &= \left\| \frac{Sx(y, \mu)}{\mu} - e \right\|^2 \\
&= \left\| \frac{S^* x^*(y, \mu)}{\mu} - e \right\|^2 + \left(\frac{s_j \xi}{\mu} - 1 \right)^2.
\end{aligned}$$

Hence

$$\delta \geq \left\| \frac{S^* x^*(y, \mu)}{\mu} - e \right\| \tag{4.31}$$

and

$$\left| \frac{\xi}{\mu} \right| \leq \frac{1 + \delta}{s_j}. \tag{4.32}$$

Now define $\Delta x := x^* - x^*(y, \mu)$. Note that

$$\begin{aligned}
b &= A^* x^* \\
&= A^* \Delta x + A^* x^*(y, \mu) \\
&= A^* \Delta x + Ax(y, \mu) - \xi a_j \\
&= A^* \Delta x + b - \xi a_j,
\end{aligned}$$

hence $A^* x^* = b$ reduces to $A^* \Delta x = \xi a_j$. Substituting this all into (4.30) we get

$$\begin{aligned}
\delta^*(y, \mu) &= \min_{\Delta x} \left\{ \left\| \frac{S^* x^*(y, \mu)}{\mu} - e + \frac{S^* \Delta x}{\mu} \right\| : A^* \Delta x = \xi a_j \right\} \\
&\leq \min_{\Delta x} \left\{ \left\| \frac{S^* x^*(y, \mu)}{\mu} - e \right\| + \left\| \frac{S^* \Delta x}{\mu} \right\| : A^* \Delta x = \xi a_j \right\}
\end{aligned}$$

$$\leq \ \delta + \min_{\Delta x}\left\{\frac{1}{\mu}\|S^*\Delta x\| \ : \ A^*\Delta x = \xi a_j\right\}$$

$$= \ \delta + \frac{\xi}{\mu}\|a_j\|_{N^*}$$

$$\leq \ \delta + \frac{1+\delta}{s_j}\|a_j\|_{N^*},$$

where the second inequality follows from (4.31) and the last equality follows since

$$\Delta x \ = \ \xi(S^*)^{-2}(A^*)^T(A^*(S^*)^{-2}(A^*)^T)^{-1}a_j$$

$$= \ \xi S_{N^*}^{-2}A_{N^*}^T(AS^{-2}A^T)_{N^*}^{-1}a_j$$

minimizes $\|S^*\Delta x\|$ over $A^*\Delta x = \xi a_j$, and the last inequality follows from (4.32). Using the definition

$$\delta_j^* = \frac{s_j}{\|a_j\|_{N^*}}$$

we conclude that

$$\delta^*(y,\mu) \leq \delta + \frac{1+\delta}{\delta_j^*}.$$

By the previous lemma, this completes the proof. □

4.2.3 The build–up and down algorithm

To keep the formulas simple we will assume in this section (as in Ye [163]) that the constraints $-e \leq y \leq e$ are included in the dual constraints $A^T y \leq c$. Let $J \subseteq N$ be the index set corresponding to the constraints $-e \leq y \leq e$. We will also assume that the columns of A have length 1. Note that each bounded problem can be formulated in this way by rescaling. At the end of this section we will show that these assumptions are not essential.

Under the above assumptions it can easily be proved (see Ye [163]), that if $Q \supseteq J$ then

$$\|a_i\|_Q^2 = a_i^T(AS^{-2}A^T)_Q^{-1}a_i \leq \frac{1}{2}. \tag{4.33}$$

Based on the analysis in the previous section we will give strategies for working with subsets of the constraints instead of the full constraint matrix A. Both a build–up strategy (adding constraints) and a build–down strategy (deleting constraints) will be given.

In our approach we start with a certain (small) subset Q of the index set N, such that $Q \supseteq J$. Then we start the logarithmic barrier method, with respect to the constraints in Q. This means that we work with the problem

$$(\mathcal{LD}_Q) \qquad \max\left\{b^T y : a_i^T y \leq c_i, \ i \in Q\right\},$$

instead of problem $(\mathcal{LD})$. If the current iterate is close to (or violates) a constraint which is not in Q, it is added to the current system, and we go back to the previous iterate. To be more precise: we check in each iteration if there is an index $i \notin Q$ such that

$$s_i < 2^{-t_a} \text{ or } s_i < 2^{-t_a}\hat{s}_i, \tag{4.34}$$

where t_a is some 'adding' parameter, and $\hat{s}$ is the slack vector of the dual iterate which was almost centered with respect to the previous value of the barrier parameter. If there is such a constraint, we add it to our system, go back to the previous iterate (for which $s_i \geq 2^{-t_a}$ and $s_i \geq 2^{-t_a}\hat{s}_i$) and continue the process. Consequently, all iterates are feasible for the whole problem. This is the build–up part.

If the slack value of a constraint in the current iterate is sufficiently large (say $s_i \geq t_d$, where t_d is a certain 'deleting' parameter), we will remove it from our current system, since it is then likely that this constraint will be nonbinding in an optimal solution. After removing constraints, we recenter when necessary. Note that it is not sure that this constraint is nonbinding in an optimal solution, but, since we have a strategy for adding constraints, this causes no problem: constraints which are incorrectly removed, will be added later on. To avoid 'cycling', i.e. the case that a constraint is deleted and added many (in fact infinitely many) times, we will only delete constraints if we are close to the central path.

Before describing the algorithm we introduce some notation. Let $\delta_Q(y,\mu)$, $\phi_B^Q(y,\mu)$ and p_Q denote the δ–measure, the barrier function and the Newton direction, respectively, with respect to the subsystem Q. Recall that computing $(AS^{-2}A^T)_Q$ costs less computational time than computing $AS^{-2}A^T$, especially when $|Q|$ is relatively small. The algorithm goes as follows.

Build–Up and Down Algorithm

Input:
ϵ is an accuracy parameter;
θ is the reduction parameter, $0 < \theta < 1$;
Q is the initial subset of constraints, $Q \supseteq J$;
μ_0 is the initial barrier parameter value;
y^0 is a given interior feasible point such that $\delta_Q(y,\mu_0) \leq \frac{1}{4}$;

begin
    $y := y_0; \hat{y} := y_0; \mu := \mu_0;$
    while $\mu > \frac{\epsilon}{4n}$ **do**
    begin
        Delete–Constraints;
        $\mu := (1-\theta)\mu;$
        Center–and–Add–Constraints
    end
end.

The procedures **Center–and–Add–Constraints** and **Delete–Constraints** are defined as follows.

Procedure Center–and–Add–Constraints

Input:
t_a is an 'adding' parameter;

begin
  while $\delta_Q(y, \mu) > \frac{1}{4}$ do
  begin
    $\tilde{\alpha} := \arg\min_{\alpha > 0} \left\{ \phi_{\mathrm{B}}^Q(y + \alpha p_Q, \mu) : s_i - \alpha a_i^T p_Q > 0, \; \forall i \in Q \right\}$;
    if $\exists i \notin Q \; : \; s_i - \tilde{\alpha} a_i^T p_Q < 2^{-t_a} \max(1, \hat{s}_i)$ then
      $Q := Q \cup \{ i : s_i - \tilde{\alpha} a_i^T p_Q < 2^{-t_a} \max(1, \hat{s}_i), i \notin Q \}$
    else $y := y + \tilde{\alpha} p_Q$
  end
  $\hat{y} := y$
end.

Procedure Delete–Constraints

Input:
$t_d \geq 4$ is a 'deleting' parameter;

begin
  for $i := 1$ to n do
    if $i \in Q \setminus J$ and $s_i \geq t_d$ then
    begin
      $Q := Q \setminus \{i\}$;
      if $\delta_Q(y, \mu) > \frac{1}{4}$ then Center–and–Add–Constraints
    end
end.

In Section 4.2.5 it is shown how to obtain a starting point y and set Q such that y is feasible for the whole problem and $\delta_Q(y, \mu) \leq \frac{1}{4}$. Moreover the minimization in the Procedure Center–and–Add–Constraints need not to be exact.

4.2.4 Complexity analysis

First we analyze the procedure Center–and–Add–Constraints. Let q be the cardinality of the current subset Q. Theorem 2.3 gives an upper bound for the number of Newton iterations required by the standard logarithmic barrier method between two reductions of the barrier parameter. Lemma 4.11 below gives a similar upper bound if also constraints are added between the two reductions. The following lemma will be needed in the proof of Lemma 4.11. It generalizes Theorem 2.3 and Lemma 4.7.

Lemma 4.10 If $\delta = \delta(y, \mu) \le \frac{1}{4}$, and $\bar{\mu} := (1 - \theta)\mu$, then

$$\phi_{\rm B}^r(y, \bar{\mu}) - \phi_{\rm B}^r(y^r(\bar{\mu}), \bar{\mu}) \le \frac{1}{3} + \frac{\theta}{(1 - \theta)^2}\left(\frac{3}{2}\sqrt{q + r} + \theta(q + r)\right) +$$

$$\sum_{k=1}^{r} \max\left(0, \ln \frac{2r\sqrt{2}}{s_{i_k}}\right),$$

where the superscript r refers to the situation that constraints $i_1, \cdots, i_r$ are added.
Proof: The proof generalizes the proof of Lemma 4.7. We construct situation 0 by
adding constraint $i_1, \cdots, i_r$ in a shifted position such that

$$s_{i_k}^0 = \max(s_{i_k}, 2r\sqrt{2}), \tag{4.35}$$

which means that

$$\delta_{i_k}^0 = \frac{s_{i_k}^0}{\|a_{i_k}\|_{N^0}}$$

$$\ge \frac{2r\sqrt{2}}{\sqrt{\frac{1}{2}}}$$

$$= 4r,$$

where $N^0 = N \cup \{i_1, \cdots, i_r\}$. Note that

$$\phi_{\rm B}^r(y, \bar{\mu}) - \phi_{\rm B}^r(y^r(\bar{\mu}), \bar{\mu}) = \phi_{\rm B}^0(y^0(\bar{\mu}), \bar{\mu}) - \phi_{\rm B}^r(y^r(\bar{\mu}), \bar{\mu})$$

$$+ \phi_{\rm B}^0(y, \bar{\mu}) - \phi_{\rm B}^0(y^0(\bar{\mu}), \bar{\mu})$$

$$+ \phi_{\rm B}^r(y, \bar{\mu}) - \phi_{\rm B}^0(y, \bar{\mu}). \tag{4.36}$$

Now we deal with the three pairs of terms in the right–hand side of (4.36) separately.

The first term $\phi_{\rm B}^0(y^0(\bar{\mu}), \bar{\mu}) - \phi_{\rm B}^r(y^r(\bar{\mu}), \bar{\mu})$ is smaller than or equal to 0, since according
to Lemma 4.4 the barrier function value in the exact center increases after shifting.

Repeatedly applying Lemma 4.5 and using $\delta_{i_k}^0 \ge 4r$, we have that

$$\delta^0(y, \mu) \le \delta + \sum_{k=1}^{r} \frac{1}{\sqrt{1 + (\delta_{i_k}^0)^2}}$$

$$\le \frac{1}{4} + r\frac{1}{\sqrt{1 + 16r^2}}$$

$$< \frac{1}{2}.$$

From (2.28) it now follows that

$$\phi_{\rm B}^0(y, \bar{\mu}) - \phi_{\rm B}^0(y^0(\bar{\mu}), \bar{\mu}) \le \frac{1}{3} + \frac{\theta}{(1 - \theta)^2}\left(\frac{3}{2}\sqrt{q + r} + \theta(q + r)\right).$$

For the third term we simply have

$$\phi_{\text{B}}^r(y,\bar{\mu}) - \phi_{\text{B}}^0(y,\bar{\mu}) = \sum_{k=1}^{r}(\ln s_{i_k}^0 - \ln s_{i_k})$$

$$= \sum_{k=1}^{r} \max\left(0, \ln \frac{2r\sqrt{2}}{s_{i_k}}\right),$$

where the last equality follows from (4.35). Substituting all this into (4.36) gives the lemma. □

Lemma 4.11 *Between two reductions of the barrier parameter, the Build–Up and Down Algorithm requires at most*

$$O\left(1 + \frac{\theta^2}{(1-\theta)^2}(q+r) + rt_a + r\ln r\right)$$

Newton iterations if r constraints were added.

Proof: Let $\bar{\mu}$ be the current value and μ the previous value of the barrier parameter (i.e. $\bar{\mu} := (1-\theta)\mu$). Let $i_1, \cdots, i_r$ denote the indices of the r constraints which are added to Q while the barrier parameter has value $\bar{\mu}$, and let y^k, for $k = 1, \cdots, r$, denote the iterate just after the i_k-th constraint has been added, whereas $y^k(\bar{\mu})$ and $\phi_{\text{B}}^k(y^k, \bar{\mu})$ denote the corresponding center and barrier function value, respectively. Finally, let y^0 denote the iterate at the moment that μ is reduced to $\bar{\mu}$ (i.e. y^0 is in the vicinity of $y^0(\mu)$), and $\phi_{\text{B}}^0(y^0,\mu)$ the corresponding barrier function value.

Since in each iteration the barrier function value decreases by a constant (Lemma 2.6), the number of Newton iterations needed to go from y^0 to the vicinity of $y^r(\bar{\mu})$ is proportional to at most

$$P := \phi_{\text{B}}^0(y^0,\bar{\mu}) - \phi_{\text{B}}^0(y^1,\bar{\mu}) + \sum_{k=1}^{r-1}\left(\phi_{\text{B}}^k(y^k,\bar{\mu}) - \phi_{\text{B}}^k(y^{k+1},\bar{\mu})\right)$$

$$+\phi_{\text{B}}^r(y^r,\bar{\mu}) - \phi_{\text{B}}^r(y^r(\bar{\mu}),\bar{\mu}). \qquad (4.37)$$

Since $\phi_{\text{B}}^k(y^k,\bar{\mu}) = \phi_{\text{B}}^{k-1}(y^k,\bar{\mu}) - \ln s_{i_k}^k$,

$$\sum_{k=1}^{r-1}\left(\phi_{\text{B}}^k(y^k,\bar{\mu}) - \phi_{\text{B}}^k(y^{k+1},\bar{\mu})\right) = \sum_{k=1}^{r-1}\left(\phi_{\text{B}}^{k-1}(y^k,\bar{\mu}) - \phi_{\text{B}}^k(y^{k+1},\bar{\mu}) - \ln s_{i_k}^k\right)$$

$$= \phi_{\text{B}}^0(y^1,\bar{\mu}) - \phi_{\text{B}}^{r-1}(y^r,\bar{\mu}) - \sum_{k=1}^{r-1}\ln s_{i_k}^k.$$

Substituting this into (4.37), while using $\phi_{\text{B}}^r(y^r,\bar{\mu}) = \phi_{\text{B}}^{r-1}(y^r,\bar{\mu}) - \ln s_{i_r}^r$, we obtain

$$P = \phi_{\text{B}}^0(y^0,\bar{\mu}) - \phi_{\text{B}}^r(y^r(\bar{\mu}),\bar{\mu}) - \sum_{k=1}^{r}\ln s_{i_k}^k$$

$$= \phi_{\text{B}}^r(y^0,\bar{\mu}) - \phi_{\text{B}}^r(y^r(\bar{\mu}),\bar{\mu}) + \sum_{k=1}^{r}\ln \frac{s_{i_k}^0}{s_{i_k}^k}. \qquad (4.38)$$

From Lemma 4.10 we have

$$\phi_{\text{B}}^r(y^0,\bar{\mu}) - \phi_{\text{B}}^r(y^r(\bar{\mu}),\bar{\mu}) \;\leq\; \frac{1}{3} + \frac{\theta}{(1-\theta)^2}\left(\frac{3}{2}\sqrt{q+r} + \theta(q+r)\right) +$$
$$\sum_{k=1}^{r}\max\left(0,\ln\frac{2r\sqrt{2}}{s_{i_k}^0}\right).$$

Consequently,

$$P \leq \frac{1}{3} + \frac{\theta}{(1-\theta)^2}\left(\frac{3}{2}\sqrt{q+r} + \theta(q+r)\right) + \sum_{k=1}^{r}\ln\frac{\max(s_{i_k}^0, 2r\sqrt{2})}{s_{i_k}^k}.$$

The lemma follows because at each stage in the algorithm we have

$$s_{i_k}^k \geq 2^{-t_a}\max(s_{i_k}^0, 1).$$

□

Let us now analyze the procedure Delete–Constraints. Lemma 4.9 gives that when a constraint is deleted in an iterate near the path, and if $s_i \geq t_d \geq 4$ then

$$\delta^*(y,\mu) \leq \frac{1}{4} + \frac{1+\frac{1}{4}}{4}\frac{1}{\sqrt{2}} < \frac{1}{2}.$$

From this point we have to recenter. As a consequence to Lemma 4.11 we obtain that recentering costs at most $O(1 + rt_a + r\ln r)$ Newton iterations, if r constraints have to be added.

Note that it is not excluded in the algorithm that a constraint is removed at each outer iteration and added during each inner loop. To avoid this undesirable behavior, several strategies can be added to the algorithm; e.g. a constraint may be deleted only once (or a fixed number of times). In the sequel K will denote the maximal number of times a constraint may be deleted. Note that K is certainly not larger than the number of updates of μ, so using Theorem 2.2

$$K \leq \frac{1}{\theta}\ln\frac{4q^*\mu^0}{\epsilon} =: K_{\max},$$

where q^* denotes the maximal number of constraints in the subsystem during the whole process.

Theorem 4.4 *After at most*

$$O\left((K+1)(q^*\ln q^* + q^*t_a) + \left(\frac{1}{\theta} + \frac{\theta q^*}{(1-\theta)^2}\right)\ln\frac{q^*\mu_0}{\epsilon}\right)$$

Newton iterations an ϵ-optimal solution has been found for $(\mathcal{LD})$, where q^* denotes the maximal number of constraints in the subsystem during the whole process.

Proof: From Theorem 2.4 it follows that the standard logarithmic barrier algorithm needs

$$O\left(\left(\frac{1}{\theta} + \frac{\theta q^*}{(1-\theta)^2}\right) \ln \frac{q^* \mu_0}{\epsilon}\right)$$

Newton iterations to obtain an ϵ–optimal solution for $(\mathcal{LD}_{Q^*})$. Since this solution is also feasible for $(\mathcal{LD})$ and the subsystem forms a relaxation of $(\mathcal{LD})$, it is also an ϵ–optimal solution for $(\mathcal{LD})$.

During the process of the algorithm at most Kq^* constraints were deleted, which costs $O(Kq^*)$ additional Newton iterations. Moreover, at most $(K+1)q^*$ constraints were added, which costs, according to Lemma 4.11, at most $O\left((K+1)(q^* t_a + q^* \ln q^*)\right)$ additional Newton iterations. Summing all Newton iterations proves the theorem. $\square$

To obtain an optimal solution, we have to take $\epsilon = 2^{-2L}$. Now, according to Theorem 4.4, we have the following result:

To obtain a 2^{-2L}–optimal solution the algorithm requires

- $O(\sqrt{q^*}L)$ Newton iterations if we take $K = O(1)$, $t_a = O\left(\frac{L}{\sqrt{n}}\right)$, and $\theta = \frac{\nu}{\sqrt{q}}$, q is the current number of constraints in our subsystem, $\nu > 0$ and independent of q, n and L;

- $O(q^*L)$ Newton iterations if we take $K = O(1)$, $t_a = O(L)$ and $0 \leq \theta \leq 1$;

- $O(q^*L \ln q^*)$ Newton iterations if we take $K = K_{\max}$ and $t_a = O(\ln q)$, where q is the current number of constraints in our subsystem, and $0 \leq \theta \leq 1$. $\square$

Note that, since it is likely that $q^* < n$, the iteration bounds for the medium- and long–step variants (the first two cases), are better than that for the standard logarithmic barrier method (Theorem 2.4).

4.2.5 Concluding remarks

Initialization

Suppose the constraints are divided into two disjoint sets Q and $\bar{Q}$ such that $Q \cup \bar{Q} = \{1, \cdots, n\}$. Now assume that y and $(x_Q, x_{\bar{Q}})$ are the (weighted) μ_0–centers for $(\mathcal{LD})$, i.e. they minimize

$$-\frac{b^T y}{\mu_0} - \sum_{i \in Q} \ln s_i - \sum_{i \in \bar{Q}} \frac{1}{\bar{q}} \ln s_i$$

over the feasible region, where $\bar{q} = |\bar{Q}|$. It is easy to verify that these centers have to satisfy

$$\begin{aligned}
A_Q x_Q + A_{\bar{Q}} x_{\bar{Q}} &= b \\
A_Q^T y + s_Q &= c_Q \\
A_{\bar{Q}}^T y + s_{\bar{Q}} &= c_{\bar{Q}}
\end{aligned}$$

$$X_Q s_Q = \mu_0 e$$
$$X_{\bar{Q}} s_{\bar{Q}} = \frac{\mu_0}{\bar{q}} e.$$

It is well–known that such points can be obtained by transforming the original problem; see e.g. Monteiro and Adler [110] and Renegar [127]. Let us introduce now an extra column $a := A_{\bar{Q}} x_{\bar{Q}}$ and an extra primal variable ξ and dual slack variable η. Then $(x_Q, 1)$ and y satisfy

$$\begin{aligned}
A_Q x_Q + \xi a &= b \\
A_Q^T y + s_Q &= c_Q \\
a^T y + \eta &= x_{\bar{Q}}^T c_{\bar{Q}} \\
X_Q s_Q &= \mu_0 e \\
\xi \eta &= \mu_0.
\end{aligned}$$

This means that y is feasible for the whole problem $(\mathcal{LD})$ and is the center for the subproblem consisting of the q constraints and an additional constraint $a^T y \leq x_{\bar{Q}}^T c_{\bar{Q}}$, which is a valid inequality for $(\mathcal{LD})$. Note that the constraints which are not in Q are condensed into this additional constraint.

General criteria

The same results can be obtained without assuming that the box constraints $-e \leq y \leq e$ are included and that the columns of A have unit length. For this purpose we have to change the criterion (4.34) for adding constraints into

$$s_i < 2^{-t_a} \max(\hat{s}_i, \|a_i\|_Q),$$

and the criterion for deleting a constraint into

$$s_i > t_d \|a_i\|_Q.$$

Increase in potential function after deleting constraints

At the end of Ye's [163] paper, it is said to be a further research topic how deleting inequalities will affect the potential function (the potential function he used is $\phi_B(y, \mu)$, with $b = 0$). The answer can be obtained very easily from Lemmas 4.9 and 2.4. Suppose that the current point is centered, i.e. $\delta(y, \mu) \leq \frac{1}{4}$, and that an inequality is removed if the corresponding slack value is larger than 4. Then according to Lemma 4.9 we have $\delta^*(y, \mu) \leq \frac{1}{2}$. Now using Lemma 2.4 we have

$$\begin{aligned}
\phi_B^*(y^*(\mu), \mu) &\geq \phi_B^*(y, \mu) - \frac{1}{3} \\
&\geq \phi_B(y, \mu) + \ln 4 - \frac{1}{3} \\
&\geq \phi_B(y(\mu), \mu) + \ln 4 - \frac{1}{3} \\
&> \phi_B(y(\mu), \mu) + 1.
\end{aligned}$$

This means that removing such a constraint increases the potential function value by at least one. Removing a constraint which is close to the current iterate will, of course, decrease the potential function value. It can easily be verified that an increase can still be guaranteed if the corresponding slack value of the removed inequality is larger than 2.16.

Deleting nonbinding constraints

We go back to the problem $(\mathcal{LD})$, and to the standard logarithmic barrier method described in Section 2.3. Ye [159], [161] showed that it is possible in IPMs to detect constraints which are binding or nonbinding in all optimal solutions.

If a nonbinding constraint is detected, it can be removed from the system. Since the central path will change by deleting constraints, the problem for the path–following analysis is how to control the number of iterations to return to the new path. Ye [159] gives a (rather awkward) solution for this problem: the detected constraints (there must be at least two) are condensed into one constraint such that the current iterate is also close to the new path. Using Lemma 4.9 we give another, more simple solution.

First note that for the detection process we need both a dual and a primal solution. According to Lemma 2.7 a primal solution $x(y, \mu)$ can easily be obtained if $\delta(y, \mu) < 1$. From Ye [161] it is easy to obtain the following lemma.

Lemma 4.12 *Let* $\delta := \delta(y, \mu) < 1$ *and* $x := x(y, \mu)$*. Then if*

$$s_i - \frac{r}{x_i}\sqrt{w_i} > 0, \tag{4.39}$$

where $r^2 = (s^T x)^2 + \|Sx\|^2$, *and* w_i *is the i–th diagonal element of the matrix*

$$XA^T(AX^2A^T)^{-1}AX,$$

then $a_i^T y \leq c_i$ *is not binding for all dual optimal solutions.* □

We will now show that δ does not change too much if a constraint is removed for which (4.39) holds.

Theorem 4.5 *If* $n \geq 6$ *and a constraint satisfies criterion (4.39) (i.e. it is non-binding in the optimum) and therefore removed from the system, then we have for the new δ–distance*

$$\delta^*(y, \mu) \leq \delta + \frac{2(1+\delta)}{n - \delta\sqrt{n}}.$$

Proof: According to Lemma 4.12 the i–th constraint can be removed if

$$s_i > \frac{r}{x_i}\sqrt{w_i} = r\sqrt{a_i^T(AX^2A^T)^{-1}a_i}. \tag{4.40}$$

Note that

$$|\frac{x_i s_i}{\mu} - 1| \leq \delta,$$

since

$$\delta = \|\frac{Sx(y,\mu)}{\mu} - e\|.$$

Consequently,

$$x_i \leq \frac{\mu}{s_i}(1+\delta).$$

Substituting this into (4.40) gives:

$$\delta_i > \frac{r}{\mu(1+\delta)}. \tag{4.41}$$

Using that $\|\frac{Xs}{\mu} - e\| = \delta$ it can easily be verified that

$$r^2 \geq \mu^2 \left((n - \delta\sqrt{n})^2 + (\sqrt{n} - \delta)^2 \right). \tag{4.42}$$

From Lemma 4.9 if follows that

$$\delta^* \leq \delta + \frac{1+\delta}{\sqrt{\delta_i^2 - 1}}. \tag{4.43}$$

Now using (4.41) and (4.42) in (4.43) it follows after some algebraic manipulations that

$$\delta^* \leq \delta + \frac{2(1+\delta)}{n - \delta\sqrt{n}},$$

if $n \geq 6$.   □

This means that the detected constraint can be removed while still staying in the quadratic convergence region of the (new) central path.

Subjects of further research

It will be a major improvement if we could prove that $q^* = \Theta(m)$ for (a variant of) the Build–Up and Down Algorithm, i.e. that at most a multiple of m constraints are in the working set. This is a subject of further research. The following observation is important in this respect[2]. Suppose that Q is the current working set, then

$$\sum_{i \in Q} \frac{1}{\delta_i^2} = \sum_{i \in Q} e_i^T S_Q^{-1} A_Q^T (AS^{-2}A^T)_Q^{-1} A_Q S_Q^{-1} e_i$$

$$= \text{trace}(I - P_{A_Q S_Q^{-1}})$$

$$= m,$$

where the last equality follows since the projection matrix $I - P_{A_Q S_Q^{-1}}$ has m eigenvalues 1, and the trace of a matrix equals its sum of the eigenvalues. Now it easily

---

[2]This observation was made by Kurt Anstreicher.

follows that, if there are $q = \gamma m$ constraints in the current working set, then for at least one of the constraints we must have

$$\delta_i \geq \sqrt{\frac{q}{m}} = \sqrt{\gamma},$$

which means that if γ is not too small then this constraint is 'far' from the current iterate and can therefore be deleted.

Another subject of further research is to generalize the approach given in this section to (non–differentiable) convex programming. Starting with an LP–relaxation of the problem, we can apply the standard logarithmic barrier method. If the iterate is close to the boundary (or out) of the feasible region, a new linear constraint can be added to the LP–relaxation. Some theoretical and computational results are reported in [20].

of all researchers that it is too small that this conjecture did not fit the observed values and therefore probable.

Another thing of fundamental interest... some on the surface...

Chapter 5

Discussion of other IPMs

In this chapter we will briefly discuss the IPMs mentioned in Section 1.3 and not treated in the previous chapters. First we will treat the path–following methods, especially the primal–dual path–following method, since the dual (primal) path–following methods are treated in the previous chapters. Then we discuss the affine scaling method, the projective and affine potential reduction methods. Following the literature, the methods are mainly described for the primal problem $(\mathcal{LP})$ (the dual problem can be treated analogously). Some open problems will also be indicated. We will not give a detailed description of all these methods. (More detailed survey papers are [45], [55], [138] and [154].) Our aim is to show that all these methods rely on some common notions: they all use the central path somehow, and the search directions are all linear combinations of two characteristic vectors. These comparisons will be carried out in the last section.

Concerning IPMs for the convex programming problem $(\mathcal{CP})$, only a few complexity results have been obtained in the literature hitherto. The results for the path–following methods for $(\mathcal{CP})$ discussed in Chapters 2 and 3 are at this moment the best.

5.1 Path–following methods

Path–following methods based on the classical logarithmic barrier function method and the center method are extensively treated in Chapters 2 and 3. Short–step path–following methods follow the central path closely, whereas the medium–, and especially the long–step methods follow the central path loosely. The analysis given in these chapters is either based on a primal or a dual method. Extensions to smooth convex programming, satisfying some conditions, are also given in the previous chapters.

Kojima et al. [78], [79] and Monteiro and Adler [110] proposed a primal–dual short–step path–following method ([79] treats the positive semi–definite linear complementarity problem). In this method steps are taken in the primal and dual space simultaneously. Recall that the Karush–Kuhn–Tucker conditions for $x(\mu)$ and $(y(\mu), s(\mu))$

are (see (2.9))

$$\begin{cases} A^T y + s = c, & s \geq 0 \\ Ax = b, & x \geq 0 \\ Xs = \mu e. \end{cases} \tag{5.1}$$

The search direction used in [78], [79] and [110] is the Newton direction for this nonlinear system of equations. Let p_x, p_y and p_s denote the search direction in the x-, y- and s-space, respectively. Then the search directions are determined by

$$\begin{pmatrix} 0 & A^T & I \\ A & 0 & 0 \\ S & 0 & X \end{pmatrix} \begin{pmatrix} p_x \\ p_y \\ p_s \end{pmatrix} = \begin{pmatrix} 0 \\ 0 \\ \mu e - Xs \end{pmatrix}. \tag{5.2}$$

These search directions are explicitly given by

$$p_x = \left[S^{-1} - S^{-1} X A^T (AS^{-1} X A^T)^{-1} AS^{-1} \right] (\mu e - Xs), \tag{5.3}$$

$$p_y = (AS^{-1} X A^T)^{-1} AS^{-1} (Xs - \mu e), \tag{5.4}$$

$$p_s = -A^T p_y. \tag{5.5}$$

We will now follow the approach given in [110]. Suppose now that (x, y, s) are approximately centered with respect to μ. Then μ is reduced by a small factor, i.e. $\mu^+ = (1 - \theta)\mu$, where $\theta = \frac{1}{10\sqrt{n}}$. It is shown in [110] that taking a full Newton step, yields points $x^+ = x + p_x$, $y^+ = y + p_y$ and $s^+ = s + p_s$, which are approximately centered with respect to μ^+. More precisely, starting with

$$\left\| \frac{Xs}{\mu} - e \right\| \leq \frac{1}{10},$$

we get

$$\left\| \frac{X^+ s^+}{\mu^+} - e \right\| \leq \frac{1}{10}.$$

Note the similarity with the δ-measure (2.12). In fact, it is easy to see that

$$\left\| \frac{Xs}{\mu} - e \right\| \geq \max \left(\delta(x, \mu), \delta(y, \mu) \right),$$

where $\delta(x, \mu)$ is the primal variant of $\delta(y, \mu)$ (see also (2.37)).

Because of the small updates in the barrier parameter, this method is a short–step method. It immediately follows from Theorem 2.2 that this primal–dual algorithm needs $O(\sqrt{n} \ln \frac{n\mu_0}{\epsilon})$ Newton iterations. This iteration bound is the same as the bound for the short–step dual (or primal) path–following methods analyzed in Sections 2.3 and 3.3. In [78] the iteration bound is $O(n \ln \frac{n\mu_0}{\epsilon})$, but this was improved to $O(\sqrt{n} \ln \frac{n\mu_0}{\epsilon})$ in [79]. All these papers also contain some tricks (rank–one updates, inexact solutions, see also Section 4.1) to obtain the overall complexity of $O(n^3 \ln \frac{n\mu_0}{\epsilon})$

arithmetic operations. Mizuno and Todd [106] analyzed a variant of Monteiro and Adler's method in which μ^+ is chosen as small as possible while maintaining the centering condition.

A natural question is now what happens if we look at larger reductions of μ, i.e. medium- or long–steps. How many iterations are now needed to reach the vicinity of the new center? Recently this open problem was solved by Jansen et al. [65]. They showed that if medium–steps are taken the iteration bound is $O(\sqrt{n}\ln\frac{n\mu_0}{\epsilon})$ and if long steps are taken the iteration bound is $O(n\ln\frac{n\mu_0}{\epsilon})$. The structure of the analysis is similar to the one given in Section 2.3. Instead of using either the primal or the dual logarithmic barrier function, they use the so–called primal–dual barrier function, which is given by

$$
\begin{aligned}
\phi_{\mathrm{B}}^{pd}(x,y,\mu) &= \phi_{\mathrm{B}}^{p}(x,\mu)+\phi_{\mathrm{B}}^{d}(y,\mu) \\
&= \frac{x^T s}{\mu} - \sum_{i=1}^{n}\ln(x_i s_i),
\end{aligned}
$$

where ϕ_{B}^{p} (2.10) and $\phi_{\mathrm{B}}^{d} = \phi_{\mathrm{B}}$ (2.8) are the primal and dual logarithmic barrier function respectively. The key result in their analysis is that in each iteration the value of this primal–dual logarithmic barrier function decreases by a certain constant.

Monteiro and Adler [111] extended their short–step analysis to convex quadratic programming, obtaining similar complexity bounds. In [112] they extended their analysis to problems with linear constraints and convex objective with an additional assumption on the third derivative (see also Appendix A). Kortanek et al. [82] studied a similar short–step path–following method for problems with convex objective and linear constraints. Each iteration, however, requires the solution of a system of nonlinear equations, which may be harder than solving the problem itself. Zhu [170], using the analysis of [82], showed that if the objective function fulfils the so–called Scaled Lipschitz Condition (see also Appendix A), then the iteration bound is $O(\sqrt{n}\ln\frac{n\mu_0}{\epsilon})$ to obtain an ϵ–optimal solution. This Scaled Lipschitz Condition holds true for e.g. linear, convex quadratic, and entropy functions. All the extensions mentioned in this paragraph are short–step methods. It should be possible to extend these results to medium– and long–step methods using the analysis in [65].

Extending the analysis of the primal–dual path–following method to (smooth) convex programming (with nonlinear constraints) seems to be difficult. Of course it is possible to apply Newton's method to the nonlinear Karush–Kuhn–Tucker conditions (2.5) for ($\mathcal{CP}$). The nonlinear equality constraints $\sum_{i=1}^{n} x_i\nabla f_i(y) = \nabla f_0(y)$, however, give troubles in the complexity analysis. One possible way to avoid these troubles is perhaps to use the symmetric conical formulation for convex programming problems, proposed by Nesterov and Nemirovsky [119].

5.2   Affine scaling methods

We first explain a version of the method for the primal problem $(\mathcal{LP})$. Starting with $(\mathcal{LP})$, the non–negativity constraints $x \geq 0$ are replaced by the ellipsoidal constraints $\|X^{-1}(\tilde{x} - x)\| < \beta \leq 1$, in which x is the current iterate and $\tilde{x}$ the variable in which the minimization is carried out. Note that the ellipsoid

$$\{\tilde{x} \; : \; A\tilde{x} = b, \; \|X^{-1}(\tilde{x} - x)\| < \beta \leq 1\}$$

is contained in the feasible region of $(\mathcal{LP})$. So, problem $(\mathcal{LP})$ is relaxed to

$$\min_{\tilde{x}} \{c^T\tilde{x} \; : \; A\tilde{x} = b, \; \|X^{-1}(\tilde{x} - x)\| < \beta \leq 1\},$$

which is easy to solve. It can easily be verified that the solution is

$$x(\beta) = x - \beta\frac{XP_{AX}Xc}{\|P_{AX}Xc\|}.$$

The search direction $-XP_{AX}Xc$ is called the *primal affine scaling direction*. Note that this search direction is simply the scaled projected gradient direction (with scaling matrix X) with respect to the objective function. The projective method (see Section 5.3) also uses the projected gradient direction but after a projective transformation instead of an affine scaling. Consequently, the affine scaling method is a natural simplification of the projective method.

Another interesting observation is that the affine scaling direction in a point on the central path is tangent to this path. This can easily be verified by differentiating (5.1) with respect to μ:

$$\begin{cases} A^Ty' + s' = 0 \\ Ax = 0 \\ Xs' + Sx' = e, \end{cases} \tag{5.6}$$

where the prime denotes the derivative with respect to μ. Now we solve x' from (5.6). Multiplying the third equation by AS^{-1} we obtain

$$AS^{-1}Xs' = AS^{-1}e,$$

and substituting $s' = -A^Ty'$ we get

$$y' = -\mu(AX^2A^T)^{-1}AS^{-1}e.$$

Hence

$$s' = A^T(AX^2A^T)^{-1}AXe,$$

and

$$x' \;\; = \;\; \frac{1}{\mu}X(I - XA^T(AX^2A^T)^{-1}AX)e$$

author	steplength	non–degeneracy assumptions
Dikin [31],[32]	$\beta = 1$	primal
Barnes [9]	$\beta < 1$	primal and dual
Vanderbei et al. [151]	long	primal and dual
Tseng and Luo [143]	$\beta = 2^{-L}$	–
Tsuchiya [147]	$\beta = \frac{1}{8}$	dual
Tsuchiya [144]	$\beta = \frac{1}{8}$	–
Gonzaga [51]	long	primal
Tsuchiya and Muramatsu [148]	long, $\gamma = \frac{2}{3}$	–

Table 5.1: Convergence proofs for affine scaling algorithms.

$$= \frac{1}{\mu^2} X(I - XA^T(AX^2A^T)^{-1}AX)Xs$$

$$= \frac{1}{\mu^2} X(I - XA^T(AX^2A^T)^{-1}AX)Xc$$

$$= \frac{1}{\mu^2} X P_{AX} Xc.$$

This shows that the continuous affine scaling path coincides with the central path if it is initiated on this path.

The affine scaling method had already been proposed by Dikin [31] in 1967. He proved convergence for the unit steplength ($\beta = 1$), under primal non–degeneracy. After Karmarkar's publication this method was rediscovered by many researchers. Barnes [9] showed that, for fixed $\beta < 1$, the method converges if both $(\mathcal{LP})$ and $(\mathcal{LD})$ are non–degenerate. Under the same non–degeneracy assumptions, Vanderbei et al. [151] allowed β to be greater than 1, as long as all components $x_i(\beta)$ remain greater than $(1 - \gamma)x_i > 0$. This corresponds to moving a proportion γ of the way to the boundary of the feasible region. This is referred to as taking long steps in the literature. Vanderbei and Lagarias [150] clarified Dikin's proof. The proof was extended to long steps by Gonzaga [51] under primal non–degeneracy.

Tseng and Luo [143] used ergodic convergence theory to show that the affine scaling method converges if a very small steplength is taken (in fact $\beta = 2^{-L}$). Tsuchiya [147] proved convergence for $\beta = 1/8$ under strong dual non–degeneracy, and later in [144] without any non–degeneracy assumptions. The last result was obtained by using a local Karmarkar potential function and projective geometry arguments. Recently, Tsuchiya and Muramatsu [148] proved global convergence for the long-step affine scaling methods using $\gamma = \frac{2}{3}$, without requiring any non–degeneracy assumption.

Adler et al. [1] and Monma and Morton[1] [107] applied the affine scaling algorithm
to the dual problem $(\mathcal{LD})$. Since the derivation of the dual affine scaling direction
given in these papers is rather cumbersome, we will give a different derivation,
analogous to the derivation of the primal affine scaling method. Starting with $(\mathcal{LD})$
the non–negativity constraints $s \geq 0$ are replaced by the ellipsoidal constraints
$\|S^{-1}(\tilde{s} - s)\| \leq \beta < 1$, where s is the slack variable of the current iterate, and $(\tilde{y}, \tilde{s})$
are the variables in which the maximization is carried out. So, problem $(\mathcal{LD})$ is
relaxed to

$$\max_{\tilde{y}} \{b^T \tilde{y} \ : \ A^T \tilde{y} + \tilde{s} = c, \ \|S^{-1}(\tilde{s} - s)\| \leq \beta < 1\}.$$

Maximizing a linear function over an ellipsoid is an easy task. It can easily be
verified that the solution is

$$y(\beta) = y + \beta \frac{(AS^{-2}A^T)^{-1}b}{\sqrt{b^T(AS^{-2}A^T)^{-1}b}}.$$

The search direction $(AS^{-2}A^T)^{-1}b$ is called the *dual affine scaling direction*.

Because of results of Megiddo and Shub [97] concerning the continuous trajectories
induced by the affine scaling direction, the primal and dual affine algorithms are
believed to be not polynomial. They showed that for the Klee–Minty cube the
continuous trajectories visit the neighborhoods of all vertices if the starting point is
chosen close to the boundary. But, since in the algorithm discrete steps are taken,
it is still an open question whether the affine scaling method is polynomial or not.
Barnes et al. [10] added 'centering steps' based on a potential function to achieve
an $O(nL)$ iteration bound. (In Section 5.5 we will show that the search directions
used by IPMs consist of an affine scaling and a centering part.)

The long–step (dual) affine scaling method has been implemented in several codes;
for computational results see e.g. [90] or [107]. In these implementations we normally
have $0.9 \leq \gamma \leq 0.9995$. Note that the convergence proof for the long–step affine
scaling methods in [148] only holds for $\gamma \leq \frac{2}{3}$. Consequently, there still remains a
gap between the theoretical and practical step sizes.

All the methods discussed above are either primal or dual methods. There also
exists a primal–dual version of the affine scaling method. The search directions
used in this method are (5.3)–(5.5) with $\mu = 0$. Taking very short steps, $\beta =
(nL \ln n)^{-1}$, Monteiro et al. [113] proved that this method has an $O(nL^2)$ iteration
bound for linear and quadratic programming. In fact, they proved that this primal–
dual algorithm is a (very) short–step path–following algorithm.

Kojima et al. [77] proposed a primal–dual affine scaling algorithm in which the
step size at each iteration is adaptively determined as the minimizing point for the
so–called primal–dual potential function (see Section 5.4). However, the derived
complexity is not polynomial. Mizuno and Nagasawa [105] obtained an $O(nL^2)$ iteration bound by choosing the step size such that the primal–dual potential function

---

[1]They credit the derivation of the dual affine scaling direction to M.J. Todd.

does not increase. Note that this complexity bound is the same as that of Monteiro et al. [113], but the step size is much larger. They also showed that an $O(\sqrt{n}L)$ iteration bound can be obtained by using centering steps when an iterate is far from the central path. Note that this is in fact the primal–dual variant of Barnes et al. [10].

The question whether or not it is possible to prove polynomiality for the primal–dual affine scaling method using a fixed percentage to the boundary as the step size, is still open. In [77] a counter example for the convergence is given, in which the step sizes in the primal and dual spaces differ, but both are fixed ratios to the boundaries.

Dikin [31] already extended the affine scaling method to convex quadratic programming. Tsuchiya [146] extended his analysis to strictly convex quadratic programming and proved global convergence under a dual non–degeneracy condition. Gonzaga and Carlos [56] proved global convergence of the affine scaling algorithm for linear equality constrained convex problems under a primal non–degeneracy assumption.

We conclude this section by deriving an affine scaling direction for the convex problem $(\mathcal{CP})$. Without loss of generality we assume that the objective function is linear, i.e. $f_0(y) = b^T y$. Starting with this problem, the constraints $f_i(y) \le 0$ are replaced by the ellipsoidal constraint $(\tilde{y} - y)^T H (\tilde{y} - y) \le \beta^2$, where y is the current iterate, and $\tilde{y}$ the variable in which the maximization is carried out, and

$$H = \sum_{i=1}^{n} \frac{\nabla^2 f_i(y)}{-f_i(y)} + \sum_{i=1}^{n} \frac{\nabla f_i(y) \nabla f_i(y)^T}{f_i(y)^2}.$$

(As in the linear case, H is the Hessian of $-\sum_{i=1}^{n} \ln(-f_i(y))$.) As a consequence of Lemma 2.20 it follows that for problems which satisfy the self–concordance condition this ellipsoid is contained in the feasible region of $(\mathcal{CP})$ if $\beta \le \frac{1}{\kappa}$, where κ is the self–concordance parameter. So, problem $(\mathcal{CP})$ is relaxed to

$$\max_{\tilde{y}} \{ b^T \tilde{y} \ : \ (\tilde{y} - y)^T H (\tilde{y} - y) \le \beta^2 \}.$$

The solution of this relaxed problem is

$$y(\beta) = y + \beta \frac{H^{-1} b}{\sqrt{b^T H^{-1} b}}.$$

Consequently, $H^{-1}b$ can be interpreted as the 'affine scaling' direction for $(\mathcal{CP})$. McCormick [92], using a different derivation, analyzed a method based on this direction. The step size is taken such that it minimizes the distance function

$$- \ln(b^T y - b^T y(\beta)) - \sum_{i=1}^{n} \ln(-f_i(y(\beta))).$$

Using this step size, McCormick [92] showed global convergence under a regularity assumption, which reduces to the non–degeneracy assumption in the linear programming case. It is still an open question whether it is possible to prove convergence if fixed step lengths are used, e.g. a fixed percentage to the boundary.

5.3   Projective potential reduction methods

For Karmarkar's [72] projective algorithm the LP problem must be in the so-called canonical form (constraints are $Ax = 0$, $e^T x = 1$, $x \geq 0$), with known optimal value. He showed that all LP problems can be transformed to this form, but this transformation is inefficient in practice. Later on Todd and Burrell [139] used a lower bound technique to remove the assumption of known optimal value. Anstreicher [2], Gay [39], Gonzaga [48] and Ye and Kojima [164] independently developed a standard form variant.

We will sketch the method for the primal problem $(\mathcal{LP})$. The auxiliary function used in this method is in fact

$$\phi_K(x, z_l) = q \ln(c^T x - z_l) - \sum_{i=1}^{n} \ln x_i,$$

where z_l is a lower bound for the (unknown) optimal value. This surrogate function is used, instead of the objective function, to measure the progress of the algorithm and is therefore called a potential function.

Note that $\phi_K(x, z_l)$ is nonconvex, however Imai [61] showed that if $q \geq n + 1$ then $e^{\phi_K(x,z_l)}$ is strictly convex. (He also showed that if $q = n$ and the feasible region is bounded, then $e^{\phi_K(x,z_l)}$ is again strictly convex.) Consequently, for fixed z_l, $\phi_K(x, z_l)$ achieves a unique minimum, which we will denote by $x(z_l)$. The Karush–Kuhn–Tucker conditions for $x(z_l)$ are

$$\begin{cases} A^T y + s = c, & s \geq 0 \\ Ax = b, & x \geq 0 \\ Xs = \frac{c^T x - z_l}{q} e. \end{cases} \qquad (5.7)$$

If we compare this with the Karush–Kuhn–Tucker conditions for the logarithmic barrier function (2.9), then we observe that $x(z_l) = x(\mu)$, for $\mu = \frac{c^T x(z_l) - z_l}{q}$, i.e. $x(z_l)$ lies on the central path[2].

A nice property of this function is that if the value of $\phi_K(x, z_l)$ is small then x is close to the optimum. More precisely: it is easy to show that if $\phi_K(x, z_l) \leq -(n + 2q)L$ and the feasible region is bounded, then $c^T x - z_l \leq 2^{-2L}$, which means that x is a 2^{-2L}-optimal solution. Consequently, if we can guarantee that the potential function value decreases at least with a certain constant Δ in each iteration, then the iteration bound is $O(\frac{1}{\Delta}(n + q)L)$ (assuming that the initial potential function value is $O((n + q)L)$).

Since $x(z_l)$ is the minimizing point for $\phi_K(x, z_l)$, we cannot guarantee a sufficient reduction in the potential function if we are 'close' to $x(z_l)$, which is on the central path. Therefore, in such a case we have to decrease the lower bound somehow, which

---

[2]In [129] it is shown that, similarly as in the logarithmic barrier and center method, $c^T x(z_l)$ is monotonically decreasing and $b^T y(z_l)$ is monotonically increasing if z_l increases.

means in fact that we define a new reference point on the central path. So, there are in fact two important ingredients: how to obtain a reduction in $\phi_K(x, z_l)$ if x is not too close to $x(z_l)$, and how to update z_l if the current point x is indeed close to $x(z_l)$. The projective method is one way to achieve this; in Section 5.4 we will discuss an alternative (affine) way.

In the projective method $q = n + 1$. Problem $(\mathcal{LP})$ is homogenized by introducing an extra variable,

$$(\mathcal{LP}') \quad \begin{cases} \min \tilde{c}(z_l)^T \tilde{x} \\ \tilde{A}\tilde{x} = 0 \\ \tilde{x}_{n+1} = 1 \\ \tilde{x} \geq 0, \end{cases}$$

where $\tilde{c}(z_l)^T = (c^T, -z_l)$, $\tilde{A} = (A, -b)$. Then $(\mathcal{LP}')$ is rescaled with $\hat{X}$, where $\hat{x}$ is the current iterate for $(\mathcal{LP}')$, i.e.

$$\hat{x} = \begin{pmatrix} x \\ 1 \end{pmatrix},$$

and x is the current iterate for $(\mathcal{LP})$. Consequently, $(\mathcal{LP}')$ is scaled into

$$(\mathcal{LP}'') \quad \begin{cases} \min \bar{c}(z_l)^T \bar{x} \\ \bar{A}\bar{x} = 0 \\ \bar{x}_{n+1} = 1 \\ \bar{x} \geq 0, \end{cases}$$

where $\bar{c}(z_l)^T = \tilde{c}(z_l)^T \hat{X}$, $\bar{A} = \tilde{A}\hat{X}$. Now the constraint $\bar{x}_{n+1} = 1$ is replaced with $e^T \bar{x} = n + 1$, for which the current $\bar{x} = e$ is central. This replacement corresponds to the next projective transformation of the feasible region:

$$\bar{x} = \frac{(n+1)\hat{X}^{-1}\tilde{x}}{e^T \hat{X}^{-1}\tilde{x}}. \tag{5.8}$$

Note that the feasible region is now contained in the unit simplex. So, finally we end up with the following problem:

$$(\mathcal{LP}''') \quad \begin{cases} \min \bar{c}(z_l)^T \bar{x} \\ \bar{A}\bar{x} = 0 \\ e^T \bar{x} = n + 1 \\ \bar{x} \geq 0. \end{cases}$$

The search direction which is used in this 'projective' space is simply the projected gradient direction for $(\mathcal{LP}''')$, i.e.

$$p = P_{\bar{B}}\bar{c}(z_l),$$

where

$$\bar{B} = \begin{bmatrix} \bar{A} \\ e^T \end{bmatrix}.$$

The next iterate in the transformed space is

$$\bar{x}(\beta) = e - \beta\frac{p}{\|p\|}.$$

We go back to the original space by rescaling and normalizing

$$x^+ = X \begin{pmatrix} \bar{x}_1(\beta) \\ \vdots \\ \bar{x}_n(\beta) \end{pmatrix} / \bar{x}_{n+1}(\beta).$$

Note that the potential function for $(\mathcal{LP}')$ is

$$\phi_{\mathrm{K}}(x, z_l) = \tilde{\phi}_{\mathrm{K}}(\tilde{x}, z_l) = (n+1)\ln\tilde{c}(z_l)^T\tilde{x} - \sum_{i=1}^{n+1}\ln\tilde{x}_i.$$

This potential function is invariant under the projective transformation (5.8). It easily follows that $\tilde{\phi}_{\mathrm{K}}(\tilde{x}, z_l)$ differs by a constant from

$$\bar{\phi}_{\mathrm{K}}(\bar{x}, z_l) = (n+1)\ln\bar{c}(z_l)^T\bar{x} - \sum_{i=1}^{n+1}\ln\bar{x}_i.$$

Hence, by decreasing $\bar{\phi}_{\mathrm{K}}$ by a constant, we will decrease ϕ_{K} in the original space by a constant.

Let us now consider the search direction p in the projective space. It can easily be verified that

$$p = \bar{c}(z_l) - \bar{A}^T\bar{y}(z_l) - \frac{\bar{c}(z_l)^Te}{n+1}e,$$

where $\bar{y}(z_l) = (\bar{A}\bar{A}^T)^{-1}\bar{A}\bar{c}(z_l)$. In terms of the original space, p can be expressed by

$$p = \begin{pmatrix} X(c - A^T\bar{y}(z_l)) \\ b^T\bar{y}(z_l) - z_l \end{pmatrix} - \frac{c^Tx - z_l}{n+1}e, \tag{5.9}$$

since $\bar{c}(z_l)^Te = c^Tx - z_l$.

From (5.9) it is easy to see that if $\|p\| \le \frac{c^Tx - z_l}{n+1}\tau$, $0 \le \tau \le 1$, then $\delta(x, \frac{c^Tx - z_l}{n+1}) \le \tau$, which means that x is close to the central path for problem $(\mathcal{LP})$. ($\|p\| = 0 \iff$

$\delta(x, \frac{c^T x - z_l}{n+1}) = 0$.) Moreover, if $\|p\| \leq \frac{c^T x - z_l}{n+1}$, then $\bar{s}(z_l) = c - A^T \bar{y}(z_l) \geq 0$, which means that $\bar{y}(z_l)$ is dual feasible. In the projective algorithm a step along p is carried out if $\bar{y}(z_l)$ is infeasible. Based on the observation that in this case $\|p\| \geq \frac{c^T x - z_l}{n+1}$, which means that x is not too close to $x(z_l)$, this leads to a reduction in the potential function of at least $\frac{1}{4}$. If $\bar{y}(z_l)$ is feasible, the lower bound z_l is increased such that one of the coordinates of $y(z_l)$ equals zero. Note that the lower bound is updated to ensure that $\|p\|$ is large enough, which will give a sufficient reduction in the potential function.

The projective algorithm described above needs $O(nL)$ iterations and $O(n^4 L)$ arithmetic operations, assuming that $m = O(n)$. Karmarkar [72] showed that this can be brought down to $O(n^{3.5}L)$ arithmetic operations by using approximate solutions and rank–one updates. The disadvantage of this analysis is that it is based on steps of fixed, short length, which does not fit in potential reduction methods. Therefore, Anstreicher [3] introduced the Goldstein–Armijo rule to safeguard the line search, and showed that the $\sqrt{n}$ reduction in the number of arithmetic operations can still be obtained.

Gonzaga [53], [48] described a different approach to the projective algorithm, namely he gave a conical interpretation of the projective algorithm. His interesting conclusion is that there is no need for problem transformation; all techniques can be applied directly to formulation $(\mathcal{LP})$. Consequently, the dimension of the problem need not to be increased. Projective algorithms have an underlying motivation based on the minimization of a zero–degree homogeneous potential function on a cone, but this cone does not have to be described.

De Ghellinck and Vial [18] proposed another primal projective algorithm. They homogenize the problem $(\mathcal{LP})$ into $(\mathcal{LP}')$. Then, using the freedom introduced by the variable x_{n+1}, they keep $\tilde{c}(z_l)^T \tilde{x}$ constant, while minimizing $\tilde{\phi}_\kappa(\tilde{x}, z_l)$. This minimization process is carried out by performing a line search along the (projected) Newton direction. The bound is updated according to a certain rule and, after updating, the line search is guaranteed to yield a reduction of the potential function value of at least a fixed quantity. Yamashita [156] proposed a dual projective algorithm, which is basically the dual variant of De Ghellinck and Vial's primal algorithm. In Section 5.5 we will see that the search directions used by all these projective methods coincide.

Roos [128] (see also Goffin and Vial [43] and Xiao and Goldfarb [155]) proposed an interesting dual projective algorithm, which is in fact also path–following. This algorithm resembles De Ghellinck and Vial's [18] algorithm, but there are some important differences: in [128] the normalizing constraint is replicated n times; a unit Newton step is taken instead of a line search; the upper bound is updated if we are close to the current reference point on the central path. This method can be characterized as a projective short–step path–following method.

Ye [160] proposed a projective algorithm which has an $O(\sqrt{n}L)$ iteration bound. Instead of introducing one extra variable to homogenize the problem, he introduced $r = \lfloor \sqrt{n} \rfloor + 1$ extra variables. Again, after the projective transformation (5.8) a step along the gradient direction in the $(n + r)$-dimensional projective space is carried out. Using a primal–dual version of the potential function (see Section 5.4), and a slightly different updating rule for the lower bound, he proved that this method has an $O(\sqrt{n}L)$ iteration bound.

Anstreicher [2] showed that fractional linear programming problems can also be solved by the projective method. Kapoor and Vaidya [71] and Ye and Tse [167] independently extended the projective method to convex quadratic programming. The number of iterations to find a 2^{-2L}-optimal solution is $O(nL)$, and each iteration requires $O(n^3 L)$ arithmetic operations. Ye and Tse [167] also gave an analysis for problems with convex objective function and linear constraints. They proved the nice result that convexity of a function is invariant under the projective transformation (5.8).

Using the conical formulation Nesterov and Nemirovsky [119] proposed a generalization of the projective method for basically the same smooth convex programming problems as given in Section 2.5 (i.e. problems for which the logarithmic barrier satisfies the self–concordance property). Under the assumption that the optimal value is known in advance, they derived an upper bound for the total number of iterations, which is comparable to the iteration bound for the long–step path–following methods.

5.4 Affine potential reduction methods

Recall Karmarkar's potential function $\phi_K(x, z_l)$ from Section 5.3. Gonzaga [49] showed that it is not necessary to do a projective transformation to obtain a reduction in the potential function value. In [49] he assumed that the optimal value is zero, while this assumption is eliminated in [52] using a lower bound technique. The search direction used is simply the scaled projected gradient direction with respect to the potential function:

$$
\begin{aligned}
p &= -X P_{AX} X \nabla \phi_K(x, z_l) \\
&= -X P_{AX} \left(\frac{q}{c^T x - z_l} X c - e \right). \quad (5.10)
\end{aligned}
$$

Using $q \geq n + \sqrt{n}$, he proved that if $\|X^{-1}p\| \geq \tau$, where $0 < \tau < 1$, then at least a constant reduction in the potential function value can be obtained by taking a step along the search direction p. If $\|X^{-1}p\| \leq \tau$ then the current iterate is close to the central path and the lower bound can be reduced to

$$
z_l^+ = c^T x - \frac{r}{q}(c^T x - z_l),
$$

where $r = n + \|X^{-1}p\|\sqrt{n}$, which also reduces the potential function. Note that if $\|X^{-1}p\| \leq \tau$, then it follows from (5.10) and (2.14) that $\delta(x, \frac{c^T x - z_l}{q}) \leq \tau$ which means that x is close to the central path.

These results immediately lead to an $O(nL)$ iteration bound. Gonzaga [52] also showed that using a smaller increase in z_l, i.e. $r = 2n$, $q = r + \nu\sqrt{n}$, $1 \leq \nu = O(1)$, this algorithm has an $O(\sqrt{n}L)$ iteration bound. For this last method one can show that two subsequent values of the lower bound correspond to two values of the barrier parameter which differ a factor $1 - \Theta(\frac{1}{\sqrt{n}})$. This means that this algorithm is in fact a medium–step algorithm.

Monteiro [108] extended Gonzaga's [52] analysis to problems with linear constraints and convex, continuously differentiable objective functions. He proved global convergence, but did not obtain an explicit upper bound for the number of iterations.

Todd and Ye [140] introduced the following primal–dual potential function[3]:

$$\phi_{\mathrm{TY}}(x, s) = q \ln(x^T s) - \sum_{i=1}^{n} \ln x_i - \sum_{i=1}^{n} \ln s_i,$$

where $q = n + \nu\sqrt{n}$, $1 \le \nu = O(1)$. Ye [162] used this function to construct an algorithm with an $O(\sqrt{n}L)$ iteration bound. (This method was the first IPM which need not to follow the central path closely, while still having an $O(\sqrt{n}L)$ bound.) Note that

$$\phi_{\mathrm{TY}}(x, s) = \phi_{\mathrm{K}}(x, b^T y) - \sum_{i=1}^{n} \ln s_i. \tag{5.11}$$

A nice feature of $\phi_{\mathrm{TY}}(x, s)$ is its symmetry in x and s. Moreover, it can be shown that if $\phi_{\mathrm{TY}}(x, s) \le -2\nu\sqrt{n}L - n\ln n$ then $x^T s \le 2^{-2L}$, which means that x is a 2^{-2L}–optimal solution. Consequently, if one can guarantee that the potential function value decreases in each iteration by at least a certain constant, it will give an $O(\sqrt{n}L)$ iteration bound (assuming that the initial potential function value is not too large).

Ye [162] achieved such a reduction. Given a dual feasible solution (y, s), a step in the primal space with steplength $\frac{0.3}{\|X^{-1}p\|}$ is carried out along the projected gradient direction with respect to $\phi_{\mathrm{TY}}(x, s)$, i.e.

$$
\begin{aligned}
p &= -X P_{AX} X \nabla \phi_{\mathrm{TY}}(x, s) \\
&= -X P_{AX} X \left(\frac{x^T s}{q} c - X^{-1} e \right). \tag{5.12}
\end{aligned}
$$

This gives a reduction in the potential value of at least 0.05 as long as $\|X^{-1}p\| \ge 0.4$. (Because of (5.11) and s is fixed, this also means that ϕ_{K} is reduced.)

Note that for fixed s the minimum of $\phi_{\mathrm{TY}}(x, s)$ is $x(z_l)$, for $z_l = b^T y$. Consequently, if the current primal iterate is too close to $x(z_l)$ we have to update our dual estimate. To be more precise: if $\|X^{-1}p\| \le 0.4$, then the dual estimate can be updated such that the potential function value reduces by at least 0.05. Again note that if $\|X^{-1}p\| \le \tau$, $0 \le \tau \le 1$, then it follows from (5.12) and (2.14) that $\delta(x, \frac{x^T s}{q}) \le \tau$. In essence the same algorithm is developed in Freund [36]. Although there is a slight

---

[3]It was recently pointed out by M.J. Todd and Y. Ye that this potential function had earlier been introduced by Tanabe [136]. Tanabe used this function to derive his algorithms, but not in a complexity analysis.

difference between the algorithms of Ye and Freund in the control sequence as to when to take a primal step and when to recompute the dual variables, the basic ideas are the same: fix s, reduce the potential function until we are close to the central path, recompute s, and so on. Note that this method is a primal method, since only steps in the primal space are taken. Freund also showed that $q = n + \sqrt{n}$ is the optimal choice from the complexity point of view.

Note that Ye's [162] potential reduction method needs $O(\sqrt{n}L)$ iterations and $O(n^{3.5}L)$ arithmetic operations. Anstreicher and Bosch [6] showed that this last figure can be brought down to $O(n^3 L)$ by using approximate solutions, rank–one updates, and a Goldstein–Armijo safeguarded line search.

Nesterov and Nemirovsky [119] generalized Ye's potential reduction method to smooth convex programming problems in the symmetric conical formulation. The iteration bound obtained is comparable to the bounds obtained for the medium– and short–step path–following methods. However, the analysis only holds true for problems which has a self–concordant logarithmic barrier, and for which the Legendre transformation of this barrier can be calculated. It appeared that for many problems this transformation is difficult to calculate. Only for linear programming, quadratically constrained convex quadratic programming and matrix norm minimization problems it has been explicitly calculated (see [119]).

Iri and Imai [63], [62] worked with a variant of ϕ_K, which they call the multiplicative function

$$\phi_M(x, z_l) = \frac{(c^T x - z_l)^q}{\prod_{i=1}^n x_i}. \tag{5.13}$$

Note that this function is closely related to ϕ_K:

$$\phi_K(x, z_l) = \ln \phi_M(x, z_l). \tag{5.14}$$

As mentioned above, Imai [61] showed that ϕ_M is convex for $q \geq n + 1$. Consequently, the Newton direction with respect to ϕ_M is well–defined. Iri and Imai [63] assumed that the optimal value is zero, and using $z_l = 0$ they do a step along the Newton direction. They only proved some convergence results, but did not give a polynomiality proof. This method was always considered as being distinct from other IPMs.

Inspired by the work of Zhang [169], Iri [62] proved that by taking a step along the Newton direction, ϕ_M can be reduced by at least a factor $\frac{7}{8}$ (which means a reduction in ϕ_K of at least $\ln \frac{8}{7}$). In this paper it is still assumed that the optimal value is zero. This result immediately leads to an $O(nL)$ iteration bound, for finding a 2^{-2L}–optimal solution.

It is likely that it is not necessary to assume that the optimal value is zero. It should be possible to use a lower bound technique, as Gonzaga did in [52], which will also make clear the similarity with Gonzaga's potential reduction method. Because of (5.14) it easily follows that $x(z_l)$, the minimizer for $\phi_K(x, z_l)$, also minimizes ϕ_M, which means that ϕ_M also parameterizes the central path. Given a lower bound z_l

the potential function $\phi_M(x, z_l)$ is minimized by means of Newton's method, until the current iterate is close to $x(z_l)$. Then the lower bound is updated, i.e. a new reference point on the central path is defined implicitly. The clue is to find a good measure for closeness to $x(z_l)$ and a good updating rule for z_l. (To our knowledge, such an analysis has not been given in the literature, perhaps also due to the fact that the Hessian of ϕ_M is rather complicated.)

Another natural idea is to use the multiplicative version of ϕ_{TY}, which we call the symmetric multiplicative function

$$\phi_{SM}(x, s) = \frac{(x^T s)^q}{\prod_{i=1}^n x_i s_i}.$$

Note that

$$\phi_{TY}(x, s) = \ln \phi_{SM}(x, s).$$

Again, given a dual (primal) feasible solution y (x), ϕ_{SM} achieves its minimal value at a point on the central path. Using a similar procedure as in Ye [162] (fix s, reduce ϕ_{SM} until the current primal iterate is close to the central path, then update the dual estimate, and so on) it should be possible to prove that a sufficient reduction in ϕ_{SM} can be obtained by doing a step along the Newton direction. Because of the symmetrical form, this will lead to an $O(\sqrt{n}L)$ iteration bound instead of $O(nL)$.

In Kojima et al. [80] a primal–dual method based on $\phi_{TY}(x, s)$ is developed for positive semi–definite linear complementarity problems. Steps are taken in the primal and dual space simultaneously. The search direction used is the Newton direction with respect to the Karush–Kuhn–Tucker conditions for $(x(\mu), y(\mu), s(\mu))$ (see (5.2)), where $\mu = \frac{x^T s}{q}$. They proved that taking a certain step along these directions gives a reduction in $\phi_{TY}(x, s)$ of at least 0.2 if $q = n + \sqrt{n}$. Consequently, this method also has an $O(\sqrt{n}L)$ iteration bound.

Note that this method can be viewed as a barrier method where the barrier parameter changes dynamically in each iteration: $\mu = \frac{x^T s}{q}$. If for example the current iterates are centered, i.e. $x = x(\mu)$ and $y = y(\mu)$, $\mu = \frac{x^T s}{n}$, then the next barrier parameter is in fact $\mu^+ = \frac{x^T s}{q}$. This case corresponds to a medium–step update, since

$$\frac{\mu^+}{\mu} = \frac{n}{q} = 1 - \frac{1}{\sqrt{n}+1}.$$

Observe that this primal–dual method needs $O(n^{3.5}L)$ arithmetic operations. Bosch and Anstreicher [14] showed that this can be reduced by a factor $\sqrt{n}$, by using approximate solutions, rank–one updates, and a primal-dual version of the Goldstein–Armijo condition.

In practical implementations it also appeared that the so–called primal–dual infeasible interior point method is very efficient. The difference with the original primal–dual potential reduction method is that all the iterates satisfy the nonnegativity constraints $x \geq 0$ and $s \geq 0$, but not necessarily the equality constraints. Kojima et al. [76] demonstrated global convergence of such a method. Their algorithm finds approximate optimal solutions if both the primal and dual problem have interior points, and detects infeasibility when the sequence of iterates diverges. Zhang [169] and Mizuno [104] proved polynomiality for this method.

The primal–dual potential reduction method shows very good practical behavior. Computational results were first given in [95] and elaborated in [16] and [88]. Mehrotra [98] invented a predictor–corrector variant, in which also second order derivatives are used. Surprisingly good results for this variant are reported in [87], which makes, up till now, this method the 'champion' of all IPMs for solving linear programs.

Monteiro [109] proved that a special variant of the primal–dual potential reduction method for linearly constrained convex problems is globally convergent. No upper bound for the number of iterations was obtained. Yamashita [157] and McCormick [93] gave a convergence analysis for the primal–dual method applied to $(\mathcal{CP})$. They proved convergence under some usual conditions, but did not obtain an upper bound for the number of iterations. Again, the dual feasibility constraints $\sum_{i=1}^{n} x_i \nabla f_i(y) = \nabla f_0(y)$ seem to cause the difficulties. It is an interesting question whether it is possible to get good complexity results for a primal–dual potential reduction method for smooth convex programming problems stated in Nesterov and Nemirovsky's [119] symmetric conical form.

5.5   Comparison of IPMs

Use of the central path

As discussed in the previous chapters and sections all IPMs use the central path implicitly or explicitly. The continuous affine scaling trajectory initiated on the central path coincides with the central path. The logarithmic barrier function, the distance function, Karmarkar's potential function, the multiplicative function and the primal–dual potential function of Todd and Ye all yield parameterizations of the central path. Both path–following and potential reduction methods use a parameter which defines a point on the central path, go to its vicinity, and define a new reference point on the central path using dual information which is available near the central path.

Path–following methods based on the logarithmic barrier function or the distance function use the central path explicitly. Starting close to a center, a new reference point on the path is defined, and the number of iterations to go to its vicinity is upper bounded. Also the number of updates in the parameter (i.e. the number of times a new reference point is defined) can be estimated.

Projective and affine potential reduction methods are based on reductions of a potential function, but use the central path implicitly. The potential function achieves its minimum (for fixed parameter) in a point on the path. A reduction in the potential function value can be guaranteed as long as the current iterate is not too close to this point on the path. If it is too close, we can use dual information which is available near the path to generate a new parameter, which means a new point on the path. The primal–dual potential reduction method is somewhat distinct: it can be viewed as a method for which the barrier parameter changes dynamically in each iteration, i.e. the reference point on the central path is shifted in each iteration[4].

Search directions

The similarities of IPMs become even more clear if we look at the search directions used in the different methods. Note that in fact each IPM is determined by its search direction (and, less important, by the choice of the step size). Especially for the projective methods it is not clear from the papers what the search direction in the original space is. In [21] we carefully calculated the search directions for all known IPMs for linear programming; see also Yamashita [156] and Gonzaga [53]. We derived the amazing result that all search directions are linear combinations of two characteristic vectors: the affine scaling and the centering direction.

The affine scaling direction is exactly the search direction used in affine scaling directions: the primal affine scaling direction (see Section 5.2) is

$$p_{aff} = -X P_{AX} X c$$

and the dual affine scaling direction is

$$d_{aff} = (A S^{-2} A^T)^{-1} b.$$

The centering direction is the Newton direction for the problem of finding the analytic center of the feasible region (see Section 3.1). More precisely, the problems for finding the analytic center for the primal and dual problem are

$$\max \{ \prod_{i=1}^{n} x_i \ : \ Ax = b, \ x \geq 0 \}$$

and

$$\max \{ \prod_{i=1}^{n} s_i \ : \ A^T y + s = c, \ s \geq 0 \},$$

respectively. Now it is easy to verify that the primal centering direction is

$$p_{cent} = X P_{AX} e,$$

and the dual centering direction is

$$d_{cent} = -(A S^{-2} A^T)^{-1} A S^{-1} e.$$

---

[4]Dik Trom may help us to clarify the similarities and differences between the IPMs. (Dik Trom is a country–boy in a famous old Dutch boy's book [74], who is always in mischief.) At some time Dik wins a dog-race by keeping a sausage in front of the dog, while sitting on the dog-cart. In path–following and potential reduction methods the 'sausage' is put somewhere on the central path and if the 'dog' is close to it, then the sausage is moved further along the path. Note that at such times the dog is close to the central path, but otherwise can be far. For path–following methods the number of times the sausage must be shifted is known and fixed, for most of the potential reduction methods this is not known. The primal–dual potential reduction method is comparable with Dik's situation: the sausage on the central path is always far enough from the dog.

cat.	author	p_{aff}	p_{cent}
1	Gill et al. [40] Gonzaga [47], [51] Roos and Vial [133], [132] Den Hertog et al. [30]	1	μ
	Renegar [127] Vaidya [149] Den Hertog et al. [22]	1	$\dfrac{\frac{(z_u - c^T x)^2}{n} - c^T p_{aff}}{z_u - c^T x + c^T p_{cent}}$
2	Barnes [9], Dikin [31] Vanderbei et al. [151] Tsuchiya and Muramatsu[148]	1	0
3	Gonzaga [53]	1	$\dfrac{c^T x - z_l - c^T p_{cent}}{b^T (AX^2 A^T)^{-1} b}$
	De Ghellinck and Vial [18] Mitchell and Todd [103] Gay [39], Ye and Kojima [164]	1	$\dfrac{c^T x - z_l - c^T p_{cent}}{1 + b^T (AX^2 A^T)^{-1} b}$
	Ye [160]	1	$\dfrac{c^T x - z_l - c^T p_{cent}}{\lfloor \sqrt{n} \rfloor + 1 + b^T (AX^2 A^T)^{-1} b}$
4	Gonzaga [49], [52]	1	$\dfrac{c^T x - z_l}{q}$
	Freund [36], Ye [162]	1	$\dfrac{x^T s}{q}$
	Iri and Imai [63]	1	$\dfrac{\frac{(c^T x - z_l)^2}{n+1} + c^T p_{aff}}{c^T x - z_l - c^T p_{cent}}$

Table 5.2: Primal search directions in 1. path–following 2. affine scaling 3. projective potential reduction 4. affine potential reduction algorithms, for linear programming.

cat.	author	d_{aff}	d_{cent}
1	Gill et al. [40] Gonzaga [47], [51] Roos and Vial [133], [132] Den Hertog et al. [30]	1	μ
	Renegar [127] Vaidya [149] Den Hertog et al. [22]	1	$\frac{\frac{(b^T y - z_l)^2}{n} + b^T d_{aff}}{b^T y - z_l - b^T d_{cent}}$
2	Barnes [9], Dikin [31] Vanderbei et al. [151] Tsuchiya and Muramatsu[148]	1	0
3	Gonzaga [54] Yamashita [156]	1	$\frac{z_u - b^T y + b^T d_{cent}}{1 + c^T S^{-1}(I - S^{-1} A^T (AS^{-2} A^T)^{-1} AS^{-1}) S^{-1} c}$
	Roos [128]	1	$\frac{z_u - b^T y + b^T d_{cent}}{n + c^T S^{-1}(I - S^{-1} A^T (AS^{-2} A^T)^{-1} AS^{-1}) S^{-1} c}$
4	Gonzaga [49] and [52]	1	$\frac{z_u - b^T y}{q}$
	Freund [36], Ye [162]	1	$\frac{x^T s}{q}$
	Iri and Imai [63]	1	$\frac{\frac{(z_u - b^T y)^2}{n+1} - b^T d_{aff}}{z_u - b^T y + b^T d_{cent}}$

Table 5.3: Dual search directions in 1. path–following 2. affine scaling 3. projective potential reduction 4. affine potential reduction algorithms, for linear programming.

cat.	author	p^*_{aff}/d^*_{aff}	p^*_{cent}/d^*_{cent}
1	Kojima et al. [78] Monteiro and Adler [110] Todd and Ye [140]	1	μ
2	Monteiro et al. [113] Mizuno and Nagasawa [105]	1	0
4	Kojima et al. [80]	1	$\frac{x^T s}{q}$

Table 5.4: Primal–dual search directions in 1. path–following 2. affine scaling 4. affine potential reduction algorithms, for linear programming.

Again note that these Newton directions coincide with the scaled projected gradient direction for the center problem. Primal–dual methods use search directions which are linear combinations of the primal–dual affine scaling and primal–dual centering direction, given by

$$p^*_{aff} = -DP_{AD}Dc,$$

$$d^*_{aff} = (AD^2A^T)^{-1}b,$$

$$p^*_{cent} = DP_{AD}DX^{-1}e,$$

$$d^*_{cent} = -(AD^2A^T)^{-1}AS^{-1}e,$$

where $D = (XS^{-1})^{1/2}$, the geometric mean of X and S^{-1}. Note that p^*_{aff} and d^*_{aff} are the projected gradient directions for $(\mathcal{LP})$ and $(\mathcal{LD})$ after scaling with D. These directions are used in the primal–dual affine scaling method of Monteiro et al. [113] and Mizuno and Nagasawa [105]. In the same way, the centering directions p^*_{cent} and d^*_{cent} are the projected gradient directions for the primal and dual center problems, respectively, after scaling with D.

The Tables 5.2–5.4 are taken from our paper [21]. They show that the search directions used by IPMs for linear programming are linear combinations of the affine scaling and centering direction (we omit the technical calculations; they can be found in [21]). Because of these results, it has been proposed (Gonzaga [53], Domich et al. [33]) to do a search on the plane spanned by the affine scaling and centering direction. Some computational results are reported in [33].

The affine scaling direction for $(\mathcal{CP})$ (with linear objective) was derived in Section 5.2, and is given by $H^{-1}b$. The centering direction is defined as the Newton direction for

$$\max \left\{ \prod_{i=1}^{n}(-f_i(y)) \; : \; f_i(y) \leq 0 \right\},$$

which is the problem of finding the analytic center of $\mathcal{F}$. This Newton direction is explicitly given by $-H^{-1}g$, where

$$g = \sum_{i=1}^{n} \frac{\nabla f_i(y)}{-f_i(y)}.$$

Again, it can be verified that both the search directions used in the logarithmic barrier method (Section 2.5) and the center method (Section 3.4) are linear combinations of these two characteristic vectors.

Complexity bounds

As we have seen, the iteration bounds for path–following and potential reduction methods are $O(nL)$ or $O(\sqrt{n}L)$. The potential function methods with an $O(\sqrt{n}L)$ iteration bound can not be guaranteed to be a real long–step method for each problem, i.e. if we look at the corresponding barrier parameters, then the reduction in the barrier parameter is not guaranteed to be $\Theta(1)$ for all problems, while for the long–step path–following methods this is possible. It can even be proved that Gonzaga's [52] potential reduction method, with an $O(\sqrt{n}L)$ iteration, is a medium–step path–following method for each problem.

A great advantage of path–following methods is that, as we have shown in Chapters 2 and 3, for the convex programming problem $(\mathcal{CP})$ similar complexity bounds as for the linear case can be obtained. As mentioned in Section 5.4 also some complexity results for the potential reduction method for convex programming have been obtained by Nesterov and Nemirovsky [119]. A disadvantage of their method is that it is necessary to compute the Legendre transformation of the barrier, which can be very difficult. Hitherto, this transformation has been calculated only for linear and convex quadratic programming with quadratic constraints and matrix norm minimization.

Practical merits

Only a relatively small number of papers on IPMs deal with implementations. Of course, it is quite easy to convert a theoretical IPM into a straightforward implementation, but it requires a great deal of work and specialized knowledge to obtain an effective implementation.

An early study of the practical behavior of the projective method was carried out by Tomlin [141]. The results obtained were not encouraging, but the main reason for this is that the reformulation used is inefficient.

After that Vanderbei et al. [151] gave experimental results for the primal affine scaling method, while Adler et al. [1], Monma and Morton [107], and Marsten et al. [90] gave extensive results for the dual affine method. Gill et al. [40] gave an implementation for the (long–step) path–following method, based on the logarithmic barrier function. The advantage of a dual variant is that the solution of the linear system of equations need not be exact, since all iterates are in the interior of the feasible region and so can absorb residual errors. All these results were promising when compared with the results of MINOS (an implementation of the simplex method).

The primal–dual potential reduction method was first implemented by McShane et al. [95]. They found that the method typically takes fewer iterations than the dual affine scaling method. The implementation was refined and extended in Choi et al. [16] and Lustig et al. [87]. They found that their implementation starts to dominate the simplex method (MINOS) when, roughly speaking, $n + m$ is greater than 2.500. After that, Mehrotra [98] developed his predictor–corrector variant of this primal–dual potential reduction method, which showed a substantial improvement in practical behavior [88].

In [89] it is argued that most of the NETLIB problems, which is a standard set of test problems, are too small to reveal the dramatic superiority of the interior method for large models. In this paper the authors showed that the speedups of their interior point implementation over the simplex implementation in OSL are significant for large problems.

At this moment there are several commercial codes which contain IPM implementations, e.g. OB1, OSL and KORBX . The OB1 code, developed by Lustig, Marsten and Shanno, contains the dual affine scaling method, the dual logarithmic barrier function method and the primal–dual potential reduction method (with and without the predictor–corrector modification). It is striking that in practice these IPMs need 20–60 iterations, almost independent on the problem dimensions. (This indicates that the theoretical worst case iteration bounds are too pessimistic.) From the practical point of view the most efficient algorithm for linear programming is the primal–dual potential reduction variant with the predictor–corrector modification.

At this time the practical merits of IPMs for (general) convex programming are still unclear. Only a few recent papers deal with interior point implementations for convex programming, but we expect that this number will increase in the near future, since now the theory has justified the practicability of some IPMs. The most important justification from the theory is that the theoretical complexity results of some IPMs for solving some classes of convex programming problems is the same as for solving linear programming problems (see Chapters 2 and 3).

In [126] an efficient implementation of the logarithmic barrier method for quadratic programming is given. Some computational results for the primal–dual method for problems with separable convex objective and linear constraints are reported in [15]. Some implementation issues for logarithmic barrier methods for general convex programming problems are treated in [153] and [115]. The computational results for the logarithmic barrier method given in [118] and [69] are promising, but still much has to be done. At this moment there are no computational results for other IPMs for general convex programming.

It is important to note that still some of the heuristics used in practice, both for linear and convex programming, are different from the theoretical implementations studied by researchers. So, in many cases there is still a gap between theory and practice.

As discussed in the Introduction, IPMs were already studied and implemented in the 1960s. A natural question is therefore why these methods did not retain their initial popularity. Several reasons can be given; see e.g. Jarre [68] and Wright [154]. The first reason is that the theoretical analysis given the last eight years, not only showed polynomiality for these methods, but also provided insight in important implementation issues. The most important insight concerns the choice of the barrier function: the complexity results given for the logarithmic barrier function could not be given for other barriers, like the inverse barrier (see [24]). Other important issues clarified by the recent literature are the choice of the step length, the usage of the central path as a reference path, the termination criteria for minimizing the barrier function, etc. The second reason is concerned with the fact that Hessians of barrier functions can become increasingly ill–conditioned as the iterates approach an optimal solution. This was the main reason why these methods became out of flavor during the 1960s. Nowadays, however, much more stable techniques are available, which can be used to overcome this difficulty. Moreover, the computers of the present day use a much higher arithmetic precision than in the 1960s. The third reason is that the efficiency of an interior point implementation heavily depends on the usage of good sparse matrix techniques, which were not available in the 1960s.

Chapter 6

Summary, conclusions and recommendations

Most of the papers on path–following methods are concerned with short–step methods. These methods are unattractive in practice since they use fixed short steps and small updates in the parameter, and therefore require many iterations. On the other hand, medium– and long–step path–following methods are much more flexible, since they allow to do large updates in the parameter and (approximate) line searches. In the literature such medium– and long–step methods are only analyzed for the logarithmic barrier method applied to linear programming.

In this book we studied short–, medium– and long–step path–following methods for linear, quadratic and smooth convex programming problems based on the classical logarithmic barrier and Huard's distance function. Both methods have much in common: both functions parameterize the so–called central path, use Newton steps to reach the vicinity of the current reference point on the central path, whereafter the parameter (barrier parameter μ or lower bound z) is updated, which means that a new reference point is defined.

We proved some important properties for the logarithmic barrier method applied to linear programming, by using the quadratic convergence of the Newton process in a well–defined vicinity of the μ–center. Based on these properties we were able to give an upper bound for the number of Newton iterations to reach the new vicinity. The number of updates in the parameter μ to obtain an ϵ–optimal solution could also be estimated. Consequently, the product of these two bounds is an upper bound for the total number of iterations.

Depending on the updating scheme we derived the following iteration bounds for obtaining an ϵ–optimal solution: $O(n \ln \frac{n\mu_0}{\epsilon})$ iterations for the long–step method and $O(\sqrt{n} \ln \frac{n\mu_0}{\epsilon})$ iterations for the medium–step method. By doing a small update in the barrier parameter it was possible to prove that only one full Newton step is sufficient to reach the vicinity of the new reference point. This result also yields an $O(\sqrt{n} \ln \frac{n\mu_0}{\epsilon})$ iteration bound for the short–step method. Although the short– and medium–step variants give the best iteration bounds, the long–step variant is much better in practice since in the long–step variant we do not have to follow the central path closely. A reason for this inconsistency might be that the $O(n)$ upper bound for the number of inner iterations for the long–step variant is rather pessimistic since the reduction in the logarithmic barrier function is usually larger than $\frac{1}{11}$ (we are allowed to do (approximate) line searches, and in many iterations we will have $\delta(y, \mu) \gg \frac{1}{2}$). On the other hand, the given iteration bounds for the short– and medium–step variants are totally determined by the number of updates in the barrier parameter, which is exact more or less.

The logarithmic barrier method was also extended to convex quadratic programming and, assuming some smoothness condition, to convex programming. In all these cases the Hessian norm of the Newton direction ($\|p\|_H$) appeared to be appropriate to measure the 'distance' to the reference point. For linear programming the analysis was facilitated by giving a useful characterization of this measure ($\delta = \|p\|_H$). A similar analysis has been given for the convex quadratic case, by using three different but related measures. The iteration bounds obtained are $O(n \ln \frac{n\mu_0}{\epsilon})$ for the long–step variant and $O(\sqrt{n} \ln \frac{n\mu_0}{\epsilon})$ for the medium– and short–step variant.

For smooth convex programming problems such a nice characterization could not be given, but using the notion of self–concordance [119] we obtained similar complexity results as for linear and quadratic programming. More precisely, the iteration bounds obtained are $O(\kappa^2 n \ln \frac{n\mu_0}{\epsilon})$ for the long–step variant and $O(\kappa \sqrt{n} \ln \frac{n\mu_0}{\epsilon})$ for the medium– and short–step variant. The complexity analysis is based on the self-concordance condition for the logarithmic barrier function. In some cases we have to reformulate the problem such that self–concordance for the resulting logarithmic barrier function can be proved. Such self–concordance proofs can be given for linear and convex quadratic programming with quadratic constraints, primal geometric programming, l_p–approximation, matrix norm minimization, maximal inscribed ellipsoid (see [119]), (extended) entropy programming, dual geometric programming, primal and dual l_p–programming (see Appendix A).

Using the observation that $y(z)$, the minimum of Huard's distance function, is the analytic center of a certain region (the intersection of the feasible region and a certain level set), it was straightforward to extend the δ–measure for this case. Based on this relationship many results for the center method for linear and convex programming directly followed from the results obtained for the logarithmic barrier method. For linear programming the following iteration bounds were obtained: $O(n \ln \frac{z^* - z^0}{\epsilon})$ for the long–step variant, and $O(\sqrt{n} \ln \frac{z^* - z^0}{\epsilon})$ for the medium– and short–step variant. For convex programming problems satisfying the self–concordance condition the following iteration bounds were obtained: $O(\kappa^2 n \ln \frac{z^* - z^0}{\epsilon})$ for the long–step variant, and $O(\kappa \sqrt{n} \ln \frac{z^* - z^0}{\epsilon})$ for the medium– and short–step variant. These bounds are similar to the iteration bounds for the logarithmic barrier method, since both $n\mu_0$ and $z^* - z^0$ indicate the initial gap in objective value.

In each iteration of the logarithmic barrier method for linear programming one has to solve a linear system like
$$AS^{-2}A^T p = r,$$
where r is some right–hand side. Solving this linear system costs $O(n^3)$ arithmetic

operations, hence the overall complexity is $O(n^4 \ln \frac{n\mu_0}{\epsilon})$ for the long–step variant and $O(n^{3.5} \ln \frac{n\mu_0}{\epsilon})$ for the medium–step variant. We analyzed two ways to cut down the work per iteration. The analysis can easily be extended to the center method, because of the close relationship between the logarithmic barrier and the center method.

First, we analyzed a variant of the logarithmic barrier method in which approximations $\tilde{S}$ for S are used. The diagonal term $\tilde{s}_j$ is updated only if it differs too much from the previous value. By performing a safeguarded line search, we proved that a limited number of components of $\tilde{S}$ are updated at a given iteration. This means that using rank–one updates a reduced computational cost can be achieved by computing $(A\tilde{S}^{-2}A^T)^{-1}$ instead of $(AS^{-2}A^T)^{-1}$. We proved that in this way a $\sqrt{n}$ reduction in the overall complexity can be obtained.

Second, we analyzed a variant of the logarithmic barrier method in which only a (promising) subset of the dual constraints are used to calculate the search direction. In particular if $n \gg m$ this can save much computational cost in computing $AS^{-2}A^T$. Starting with a certain subset of the dual constraints, we apply the standard path–following algorithm, until the current iterate is close to a constraint which is not in the current subset. Then this (almost violated) constraint is added. Also a strategy to delete constraints was given. By proving some basic lemmas concerning the effects of shifting, adding and deleting constraints on the δ–measure and the logarithmic barrier function, we were able to give upper bounds for the number of iterations. These bounds are comparable with those obtained for the standard logarithmic barrier method, with the exception that the parameter n is replaced by q^*, which is the highest number of constraints in the subsystem. It is still an open question whether it is possible to modify the build–up and down algorithm such that $q^* = \Theta(m)$ can be proved.

We also sketched the place of these path–following methods within other IPMs. Basically, there are four classes: path–following methods, affine scaling methods, projective potential reduction methods and affine potential reduction methods. The affine scaling method is believed not to be polynomial. Both projective and affine potential reduction methods use the central path implicitly. The basic idea behind these methods is the same as for the path–following methods: define a reference point on the central path, try to come to its vicinity, if the current iterate is 'too close', then update the parameter (lower or upper bound, barrier parameter, duality gap), which means that the reference point is shifted. The difference is that the progress of potential reduction methods is measured by a potential function. Another striking similarity between IPMs concerns the search direction. It appeared that the search directions used in IPMs are all linear combinations of two characteristic vectors: the affine scaling and the centering direction.

It is an important question why the logarithmic function plays such a key role in all polynomial IPMs. This can partly be explained by some nice properties of this function (symmetric, self–concordant). However, the question whether it is possible to prove polynomiality for other interior barrier methods or exterior penalty methods, is still not fully answered. We studied inverse barrier methods for linear programming in [26], but could not prove polynomiality.

We also argued that the path–following methods analyzed in this book are most suitable of all IPMs to be extended to smooth convex programming. Most of the papers on potential reduction methods for convex programming only deal with global convergence properties. Some difficulties arise when trying to analyze (affine, projective) potential reduction methods for convex programming: the dual feasibility constraints $\sum_{i=1}^n x_i \nabla f_i(y) = \nabla f_0(y)$ are difficult to satisfy, potential functions are usually nonconvex.

Nesterov and Nemirovsky [119] circumvented these difficulties by transforming the convex programming problem into a symmetric conical formulation. Using this formulation they obtained some complexity results for both a projective and an affine potential reduction method. However, these results only hold if the Legendre transformation of the (logarithmic) barrier can be explicitly given. Hitherto, this has only been accomplished for linear and convex quadratic programming with quadratic constraints and for matrix norm minimization problems.

Based on the theoretical results, several efficient IPM implementations have been developed, all using good sparse matrix techniques. Numerical results [91] have shown that interior methods start to dominate the simplex method when $m + n$ is greater than 2.500. For large linear programming problems IPMs are much faster than the simplex method. Especially some variants of the primal–dual potential reduction variant have shown very good practical behavior. Again we note that since the heuristics used in practical implementations often differ from the theoretical algorithms, there are still a lot of open theoretical problems.

It has been observed that in practice IPMs need 20–60 iterations, almost independent of the problem dimensions. It is an intriguing open problem to give a theoretical explanation for this phenomenon. Another closely related question is whether it is possible to improve the best iteration bound $(O(\sqrt{n}L))$ and the best overall complexity $(O(n^3L))$.

Concerning convex programming, the conversion of theoretical results into numerical algorithms has been slow so far. Since the complexity analysis and results for path–following methods for linear and convex programming are similar, this may indicate that it is possible to develop efficient implementations for path–following methods for convex programming based on the theoretical results. Some preliminary implementations have provided some encouragement that IPMs are in practice efficient for solving convex programming problems, but still much has to be done in this respect (e.g. the use of good sparse matrix techniques). The analysis given in Chapters 2 and 3 may help to obtain an efficient IPM implementation for convex programming.

Another challenge is to carry these new interior point methodologies into other classical areas of mathematical programming, e.g sensitivity and parametric analysis, (mixed) integer programming, cutting plane and decomposition methods, etc. Until recently, it was believed by many researchers that sensitivity and parametric analysis could not be handled with IPMs. In [64], however, it is shown that many difficulties arising from the simplex approach are circumvented by the interior point approach. A typical feature of an IPM is that it quickly finds feasible solutions with good objective values, but takes a relatively long time to converge to an accurate solution. This feature can be of great value for branch and bound methods for solving (mixed) integer programming (see [13]). It is well-known that many classical cutting plane and decomposition methods are unstable and inefficient in practice. It is not unlikely that using 'central interior points' rather than 'boundary points' will put new life into these methods (see [42]).

Appendix A

Self–concordance proofs

In this appendix we show that for some classes of problems the logarithmic barrier function and the distance function are self–concordant. Moreover, we show that some other smoothness conditions used in the literature are also covered by this self–concordance condition.

Recall the definition of self–concordance as given in Section 2.5.2:

Definition of self–concordance: A function φ : $\mathcal{F}^0 \to \mathbb{R}$ is called κ–self–concordant on $\mathcal{F}^0$, $\kappa \geq 0$, if φ is three times continuously differentiable in $\mathcal{F}^0$ and for all $y \in \mathcal{F}^0$ and $h \in \mathbb{R}^m$ the following inequality holds:

$$|\nabla^3 \varphi(y)[h, h, h]| \leq 2\kappa \left(h^T \nabla^2 \varphi(y) h \right)^{\frac{3}{2}},$$

where $\nabla^3 \varphi(y)[h, h, h]$ denotes the third differential of φ at y and h, i.e.

$$\nabla^3 \varphi(y)[h, h, h] = \sum_i \sum_j \sum_k \frac{\partial^3 \varphi(y)}{\partial y_i \, \partial y_j \, \partial y_k} h_i h_j h_k.$$

A.1 Some general composition rules

The following lemma gives some helpful composition rules for self–concordant functions. The proof follows immediately from the definition of self–concordance.

Lemma A.1 *(Nesterov and Nemirovsky [119])*

- *(addition and scaling) Let φ_i be κ_i–self–concordant on $\mathcal{F}_i^0$, $i = 1, 2$, and $\rho_1, \rho_2 \in \mathbb{R}^+$ then $\rho_1 \varphi_1 + \rho_2 \varphi_2$ is κ–self–concordant on $\mathcal{F}_1^0 \cap \mathcal{F}_2^0$, where $\kappa = \max\{\frac{\kappa_1}{\sqrt{\rho_1}}, \frac{\kappa_2}{\sqrt{\rho_2}}\}$.*

- *(affine invariance) Let φ be κ–self–concordant on $\mathcal{F}^0$ and let $\mathcal{B}(y) = By + b :$ $\mathbb{R}^k \to \mathbb{R}^m$ be an affine mapping such that $\mathcal{B}(\mathbb{R}^k) \cap \mathcal{F}^0 \neq \emptyset$. Then $\varphi(\mathcal{B}(.))$ is κ–self–concordant on $\{y : \mathcal{B}(y) \in \mathcal{F}^0\}$.* $\square$

The next lemma states that if the quotient of the third and second order derivative of $f(x)$ is bounded by the second order derivative of $-\sum_{i=1}^{n} \ln x_i$, then the corresponding logarithmic barrier and distance functions are self–concordant. This lemma will help to simplify self–concordance proofs in the sequel.

Lemma A.2 *Let $f(x) \in C^3(\mathcal{F}^0)$ and convex. If there exists a β such that*

$$|\nabla^3 f(x)[h, h, h]| \leq \beta h^T \nabla^2 f(x) h \sqrt{\sum_{i=1}^{n} \frac{h_i^2}{x_i^2}}, \qquad (A.1)$$

$\forall x \in \mathcal{F}^0$ *and* $\forall h \in \mathbb{R}^n$, *then*

$$\varphi(x) := \frac{f(x)}{\mu} - \sum_{i=1}^{n} \ln x_i,$$

with $\mu > 0$, is $(1 + \frac{1}{3}\beta)$–self-concordant on $\mathcal{F}^0$, and

$$\psi(t, x) := -q \ln(t - f(x)) - \sum_{i=1}^{n} \ln x_i,$$

with $q \geq 1$, is $(1 + \frac{1}{3}\beta)$–self-concordant on $\mathbb{R} \times \mathcal{F}^0$.

Proof: We start by proving the first part of the lemma. Note that since (A.1) is scale independent, we may assume that $\mu = 1$. Straightforward calculations yield

$$\nabla\varphi(x)^T h \;=\; \nabla f(x)^T h - \sum_{i=1}^{n} \frac{h_i}{x_i} \qquad (A.2)$$

$$h^T \nabla^2 \varphi(x) h \;=\; h^T \nabla^2 f(x) h + \sum_{i=1}^{n} \frac{h_i^2}{x_i^2} \qquad (A.3)$$

$$\nabla^3 \varphi(x)[h, h, h] \;=\; \nabla^3 f(x)[h, h, h] - 2 \sum_{i=1}^{n} \frac{h_i^3}{x_i^3}. \qquad (A.4)$$

We show that

$$(\nabla^3 \varphi(x)[h, h, h])^2 \leq 4(1 + \frac{1}{3}\beta)^2 (h^T \nabla^2 \varphi(x) h)^3, \qquad (A.5)$$

from which the lemma follows. Since f is convex, the two terms on the right–hand side of (A.3) are nonnegative, i.e. the right–hand side can be abbreviated by

$$h^T \nabla^2 \varphi(x) h = a^2 + b^2, \qquad (A.6)$$

with $a, b \geq 0$. Because of (A.1) we have that

$$|\nabla^3 f(x)[h, h, h]| \leq \beta a^2 b.$$

Obviously

$$\left| \sum_{i=1}^{n} \frac{h_i^3}{x_i^3} \right| \leq \sum_{i=1}^{n} \frac{h_i^2}{x_i^2} \sqrt{\sum_{i=1}^{n} \frac{h_i^2}{x_i^2}} = b^3.$$

So we can bound the right–hand side of (A.4) by

$$|\nabla^3 \varphi(x)[h, h, h]| \leq \beta a^2 b + 2b^3. \tag{A.7}$$

It is straightforward to verify that

$$(\beta a^2 b + 2b^3)^2 \leq 4(1 + \frac{1}{3}\beta)^2 (a^2 + b^2)^3.$$

Together with (A.6) and (A.7) our claim (A.5) follows and hence the first part of the lemma.

Now we prove the second part of the lemma. Let

$$\tilde{x} = \begin{pmatrix} t \\ x \end{pmatrix}, \quad h = \begin{pmatrix} h_0 \\ \vdots \\ h_n \end{pmatrix} \quad \text{and} \quad g(\tilde{x}) = t - f(x), \tag{A.8}$$

then

$$\psi(\tilde{x}) = -q \ln g(\tilde{x}) - \sum_{i=1}^{n} \ln x_i \tag{A.9}$$

$$\nabla \psi(\tilde{x})^T h = -q \frac{\nabla g(\tilde{x})^T h}{g(\tilde{x})} - \sum_{i=1}^{n} \frac{h_i}{x_i} \tag{A.10}$$

$$h^T \nabla^2 \psi(\tilde{x}) h = -q \frac{h^T \nabla^2 g(\tilde{x}) h}{g(\tilde{x})} + q \frac{(\nabla g(\tilde{x})^T h)^2}{g(\tilde{x})^2} + \sum_{i=1}^{n} \frac{h_i^2}{x_i^2} \tag{A.11}$$

$$\nabla^3 \psi(\tilde{x})[h, h, h] = -q \frac{\nabla^3 g(\tilde{x})[h, h, h]}{g(\tilde{x})} + 3q \frac{(h^T \nabla^2 g(\tilde{x}) h) \nabla g(\tilde{x})^T h}{g(\tilde{x})^2}$$
$$-2q \frac{(\nabla g(\tilde{x})^T h)^3}{g(\tilde{x})^3} - 2 \sum_{i=1}^{n} \frac{h_i^3}{x_i^3}. \tag{A.12}$$

We show that

$$(\nabla^3 \psi(\tilde{x})[h, h, h])^2 \leq 4(1 + \frac{1}{3}\beta)^2 (h^T \nabla^2 \psi(\tilde{x}) h)^3, \tag{A.13}$$

which will prove the lemma. Since g is concave, all three terms on the right–hand side of (A.11) are nonnegative, i.e. the right–hand side can be abbreviated by

$$h^T \nabla^2 \psi(\tilde{x}) h = a^2 + b^2 + c^2, \tag{A.14}$$

with $a, b, c \geq 0$. Due to (A.1) we have

$$\left| \frac{\nabla^3 g(\tilde{x})[h, h, h]}{g(\tilde{x})} \right| \leq \beta a^2 c,$$

so that we can bound the right–hand side of (A.12) by

$$|\nabla^3 \psi(\tilde{x})[h, h, h]| \leq \beta a^2 c + 3a^2 b + 2b^3 + 2c^3. \tag{A.15}$$

It is straightforward to verify that

$$(\beta a^2 c + 3a^2 b + 2b^3 + 2c^3)^2 \le 4(1 + \frac{1}{3}\beta)^2 (a^2 + b^2 + c^2)^3,$$

by eliminating all odd powers in the second term via inequalities of the type $2ab \le a^2 + b^2$. Together with (A.14) and (A.15) our claim (A.13) follows, and hence the lemma. □

In the next sections we will show self–concordance for the logarithmic barrier function for several nonlinear programming problems by showing that (A.1) is fulfilled. An immediate consequence from Lemma A.2 is that also the distance functions for these problems are self–concordant. Note that due to Lemma A.1 also 'mixtures' of these problems have self–concordant barrier and distance functions. Finally, recall from Section 2.5.2 that $-\ln(-f_i(y))$ with $f_i(y)$ a linear or convex quadratic function, is 1–self–concordant. Consequently, in the sequel of this Appendix we will only concentrate on non–linear and non–quadratic functions $f_i(y)$.

A.2   The dual geometric programming problem

Let $\{I_k\}_{k=1,\ldots,r}$ be a partition of $\{1,\ldots,n\}$ (i.e. $\cup_{k=1}^r I_k = \{1,\ldots,n\}$ and $I_k \cap I_l = \emptyset$ for $k \ne l$). The dual geometric programming problem is then given by (see [34])

$$(\mathcal{DGP}) \quad \begin{cases} \min \ c^T x + \sum_{k=1}^r \left[\sum_{i \in I_k} x_i \ln x_i - \left(\sum_{i \in I_k} x_i \right) \ln \left(\sum_{i \in I_k} x_i \right) \right] \\ Ax = b \\ x \ge 0. \end{cases}$$

For this problem we have the following lemma.

Lemma A.3 *The logarithmic barrier function of the dual geometric programming problem $(\mathcal{DGP})$ is 2–self–concordant[1].*

Proof: Because of Lemma A.1, it suffices to verify 2–self–concordance for the following logarithmic barrier function

$$\varphi(x) = \sum_{i \in I_k} x_i \ln x_i - \left(\sum_{i \in I_k} x_i \right) \ln \left(\sum_{i \in I_k} x_i \right) - \sum_{i \in I_k} \ln x_i, \qquad (A.16)$$

for some fixed k. For simplicity, we will drop the subscript $i \in I_k$. Now we can use Lemma A.2, so that we only have to verify (A.1) for

$$f(x) := \sum x_i \ln x_i - \left(\sum x_i \right) \ln \left(\sum x_i \right),$$

---

[1]This contradicts the remark in Kortanek and No [81] that the self–concordance property does not hold for this problem.

and $\beta = 3$, which is equivalent to the following inequality:

$$\left| \sum \frac{h_i^3}{x_i^2} - \frac{(\sum h_i)^3}{(\sum x_i)^2} \right| \leq 3 \left(\sum \frac{h_i^2}{x_i} - \frac{(\sum h_i)^2}{\sum x_i} \right) \sqrt{\sum \frac{h_i^2}{x_i^2}}. \qquad (A.17)$$

Here $x_i > 0$ and h_i arbitrary. To prove this inequality, let us define:

$$\xi_i = x_i^{-\frac{1}{2}} (h_i - \frac{\sum h_j}{\sum x_j} x_i).$$

Note that

$$\sum x_i^{\frac{1}{2}} \xi_i = 0.$$

Using this substitution, we can rewrite the left–hand side of the inequality (A.17):

$$\begin{aligned}
\left| \sum \frac{h_i^3}{x_i^2} - \frac{(\sum h_i)^3}{(\sum x_i)^2} \right| &= \left| \sum \left(x_i^{-\frac{1}{2}} \xi_i^3 + 3\xi_i^2 \frac{\sum h_j}{\sum x_j} \right) \right| \\
&= \left| \sum \xi_i^2 \left(x_i^{-\frac{1}{2}} \xi_i + 3 \frac{\sum h_j}{\sum x_j} \right) \right| \\
&= \left| \sum \xi_i^2 \left(\frac{h_i}{x_i} - \frac{\sum h_j}{\sum x_j} + 3 \frac{\sum h_j}{\sum x_j} \right) \right| \\
&= \left| \sum \xi_i^2 \left(\frac{h_i}{x_i} + 2 \frac{\sum h_j}{\sum x_j} \right) \right| \\
&\leq \sum \xi_i^2 \frac{|h_i|}{x_i} + 2 \sum \xi_i^2 \frac{|\sum h_j|}{\sum x_j} \\
&\leq 3 \sum \xi_i^2 \sqrt{\sum \frac{h_j^2}{x_j^2}}, \qquad (A.18)
\end{aligned}$$

where the last inequality follows because

$$\frac{|h_i|}{x_i} \leq \sqrt{\sum \frac{h_j^2}{x_j^2}},$$

and

$$\sum \frac{h_i^2}{x_i^2} \left(\sum x_i \right)^2 \geq \sum \frac{h_i^2}{x_i^2} \sum x_i^2 \geq \left(\sum |h_i| \right)^2. \qquad (A.19)$$

(The last inequality in (A.19) follows directly from the Cauchy–Schwartz inequality.)
Now note that the right–hand side of (A.17) is equal to

$$3 \sum \xi_i^2 \sqrt{\sum \frac{h_i^2}{x_i^2}}.$$

Together with (A.18), this completes the proof. $\qquad\qquad\qquad \square$

A.3   The extended entropy programming problem

The extended entropy programming problem is defined as

$$(\mathcal{EEP}) \quad \begin{cases} \min\ c^T x + \sum_{i=1}^n f_i(x_i) \\ Ax = b \\ x \geq 0, \end{cases}$$

where $|f_i'''(x_i)| \leq \kappa_i \frac{f_i''(x_i)}{x_i}$. This class of problems is studied in Ye and Potra [166] and Han et al[2]. [59]. In the case of entropy programming we have $f_i(x_i) = x_i \ln x_i$, for all i, and $\kappa_i = 1$. Self–concordance for the logarithmic barrier function of this problem simply follows from the following lemma.

Lemma A.4 *Suppose that $|f_i'''(x_i)| \leq \kappa_i \frac{f_i''(x_i)}{x_i}$, then the logarithmic barrier function for the extended entropy programming problem $(\mathcal{EEP})$ is $(1 + \frac{1}{3}\max_i \kappa_i)$–self–concordant.*

Proof: Using Lemma A.1 it suffices to show that

$$f_i(x_i) - \ln x_i$$

is $(1 + \frac{1}{3}\kappa_i)$–self–concordant. Since

$$|f_i'''(x_i)| \leq \kappa_i f_i''(x_i) \frac{1}{x_i},$$

this immediately follows from Lemma A.2.   □

A.4   The primal l_p–programming problem

Let $p_i \geq 1$, $i = 1, \cdots, n$. Let I_k, $k = 1, \cdots, r$ be sets of indices such that $I_k \cap I_j = \emptyset, j \neq k$, and $\cup_{k=1}^r I_k = \{1, \cdots, n\}$. Then the primal l_p–programming problem [122]–[124], [137] can be formulated as

$$(\mathcal{PL}_p) \quad \begin{cases} \max\ \eta^T y \\ \sum_{i \in I_k} \frac{1}{p_i} |a_i^T y - c_i|^{p_i} + b_k^T y - d_k \leq 0, \ k = 1, \cdots, r. \end{cases}$$

Nesterov and Nemirovsky [119] treated a special case of this problem, namely the so-called l_p–approximation problem. We will reformulate $(\mathcal{PL}_p)$ such that all problem functions remain convex, contrary to Nesterov and Nemirovsky's reformulation.

---

[2]In this paper it is conjectured that these problems do not satisfy the self–concordance condition. The lemma shows that it does satisfy the self–concordance condition.

The primal l_p-programming problem can be reformulated as:

$$
(\mathcal{PL}'_p) \quad
\begin{cases}
\max \ \eta^T y \\[2mm]
\sum_{i \in I_k} \frac{1}{p_i} t_i + b_k^T y - d_k \leq 0, \ k = 1, \cdots, r \\[2mm]
\left. \begin{array}{l}
s_i^{p_i} \leq t_i \\[1mm]
a_i^T y - c_i \leq s_i \\[1mm]
-a_i^T y + c_i \leq s_i
\end{array} \right\} \ i = 1, \cdots, n \\[4mm]
s \geq 0.
\end{cases}
\qquad (A.20)
$$

The logarithmic barrier function for this problem can be proved to be self–concordant. Observe that in the transformed problem we have $4n + r$ constraints, compared with r in the original problem $(\mathcal{PL}_p)$.

Lemma A.5 *The logarithmic barrier function for the reformulated l_p–programming $(\mathcal{PL}'_p)$ problem is $(1 + \frac{1}{3} \max_i |p_i - 2|)$–self–concordant.*

Proof: Since $f(s_i) := s_i^{p_i}$, $p_i \geq 1$, satisfies (A.1) with $\beta = |p_i - 2|$, we have from Lemma A.2 that

$$- \ln(t_i - s_i^{p_i}) - \ln s_i$$

is $(1 + \frac{1}{3}|p_i - 2|)$–self–concordant. Consequently, it follows from Lemma A.1 that the logarithmic barrier function for the reformulated primal l_p–programming problem is $(1 + \frac{1}{3} \max_i |p_i - 2|)$–self–concordant. $\qquad \square$

Note that the concordance parameter depends on p_i. We can improve it as follows. We replace the constraints $s_i^{p_i} \leq t_i$ by the equivalent constraints $s_i \leq t_i^{\pi_i}$, where $\pi_i = \frac{1}{p_i}$. Moreover, the redundant constraints $s \geq 0$ are replaced by $t \geq 0$. So, we obtain the following reformulated l_p–programming problem:

$$
(\mathcal{PL}''_p) \quad
\begin{cases}
\max \ \eta^T y \\[2mm]
\sum_{i \in I_k} \frac{1}{p_i} t_i + b_k^T y - d_k \leq 0, \ k = 1, \cdots, r \\[2mm]
\left. \begin{array}{l}
s_i \leq t_i^{\pi_i} \\[1mm]
a_i^T y - c_i \leq s_i \\[1mm]
-a_i^T y + c_i \leq s_i
\end{array} \right\} \ i = 1, \cdots, n \\[4mm]
t \geq 0.
\end{cases}
\qquad (A.21)
$$

The following lemma improves the result of Lemma A.5.

Lemma A.6 *The logarithmic barrier function for the reformulated l_p–programming problem $(\mathcal{PL}''_p)$ is $\frac{5}{3}$–self–concordant.*

Proof: Since $f(t_i) := -t_i^{\pi_i}$, $\pi_i \leq 1$, satisfies (A.1) with $\beta = |\pi_i - 2|$, we have from Lemma A.2 that

$$- \ln(t_i^{\pi_i} - s_i) - \ln t_i$$

is $(1 + \frac{1}{3}|\pi_i - 2|)$–self–concordant, where $\pi_i \leq 1$. Consequently, the corresponding logarithmic barrier function is $\frac{5}{3}$–self–concordant.   □

A.5   The dual l_p–programming problem

Let q_i be such that $\frac{1}{p_i} + \frac{1}{q_i} = 1$. Moreover, the columns of matrix A are a_i, $i = 1, \cdots, n$, and the columns of matrix B are b_k, $k = 1, \cdots, r$. Then, the dual of the l_p–programming problem $(\mathcal{PL}_p)$ is (see [122]–[124], [137])

$$(\mathcal{DL}_p) \quad \begin{cases} \min \ c^T x + d^T z + \sum_{k=1}^r z_k \sum_{i \in I_k} \frac{1}{q_i} \left| \frac{x_i}{z_k} \right|^{q_i} \\ Ax + Bz = \eta \\ z \geq 0. \end{cases}$$

(If $x_i \neq 0$ and $z_k = 0$, then $z_k \left| \frac{x_i}{z_k} \right|^q$ is defined as ∞.) The above problem is equivalent to

$$(\mathcal{DL}'_p) \quad \begin{cases} \min \ c^T x + d^T z + \sum_{i=1}^n \frac{1}{q_i} t_i \\ s_i^{q_i} z_k^{-q_i+1} \leq t_i, \ i \in I_k, \ k = 1, \cdots, r \\ x \leq s \\ -x \leq s \\ Ax + Bz = \eta \\ z \geq 0 \\ s \geq 0. \end{cases} \qquad (A.22)$$

Note that the original problem $(\mathcal{DL}_p)$ has r inequalities, and the reformulated problem $(\mathcal{DL}'_p)$ $4n + r$. Now we prove the following lemma.

Lemma A.7 *The logarithmic barrier function of the reformulated dual l_p–programming problem $(\mathcal{DL}'_p)$ is $(1 + \frac{\sqrt{2}}{3} \max_i (q_i + 1))$–self–concordant.*

Proof: It suffices to show that

$$- \ln(t_i - s_i^{q_i} z_k^{-q_i+1}) - \ln z_k - \ln s_i$$

is $(1 + \frac{\sqrt{2}}{3}(q_i + 1))$–self–concordant, or equivalently by Lemma A.2, that (we will omit the subscript i and k in the sequel of this proof) $f(s, z) := s^q z^{-q+1}$ satisfies (A.1) for $\beta = \sqrt{2}(q + 1)$, i.e. that

$$|\nabla^3 f(s, z)[h, h, h]| \leq \sqrt{2}(q + 1) h^T \nabla^2 f(s, z) h \sqrt{\frac{|h_1|^2}{s^2} + \frac{|h_2|^2}{z^2}}, \qquad (A.23)$$

where $h^T = (h_1, h_2)$. After doing some straightforward calculations we obtain for the second order term

$$
\begin{aligned}
h^T \nabla^2 f(s, z) h &= q(q-1)s^{q-3}z^{-q-2}(sz^3 h_1^2 + s^3 z h_2^2 - 2s^2 z^2 h_1 h_2) \\
&= q(q-1)s^{q-3}z^{-q-2}(zh_1 - sh_2)^2 sz,
\end{aligned}
$$

and for the third order term

$$
\begin{aligned}
|\nabla^3 f(s, z)[h, h, h]| &= q(q-1)s^{q-3}z^{-q-2}|(q-2)z^3 h_1^3 - (q+1)s^3 h_2^3 - \\
& \quad 3(q-1)sz^2 h_1^2 h_2 + 3qs^2 z h_1 h_2^2| \\
&= q(q-1)s^{q-3}z^{-q-2}(zh_1 - sh_2)^2|(q-2)zh_1 - (q+1)sh_2| \\
&\leq q(q-1)(q+1)s^{q-3}z^{-q-2}(zh_1 - sh_2)^2(z|h_1| + s|h_2|).
\end{aligned}
$$

Now we obtain

$$
\frac{|\nabla^3 f(s, z)[h, h, h]|}{h^T \nabla^2 f(s, z) h} \leq (q+1) \left(\frac{|h_1|}{s} + \frac{|h_2|}{z} \right) \leq \sqrt{2}(q+1) \sqrt{\frac{|h_1|^2}{s^2} + \frac{|h_2|^2}{z^2}}.
$$

This proves (A.23) and hence the lemma. $\qquad\square$

We can improve this result as follows: the constraints $s_i^{q_i} z_k^{-q_i+1} \leq t_i$ are replaced by the equivalent constraints $t_i^{\rho_i} z_k^{-\rho_i+1} \geq s_i$, where $\rho_i := \frac{1}{q_i}$, and the redundant constraints $s \geq 0$ are replaced by $t \geq 0$. The new reformulated dual l_p-programming problem becomes:

$$
(\mathcal{DL}_p'') \quad
\begin{cases}
\min \; c^T x + d^T z + \sum_{i=1}^n \frac{1}{q_i} t_i \\[4pt]
s_i \leq t_i^{\rho_i} z_k^{-\rho_i+1}, \; i \in I_k, \; k = 1, \cdots, r \\[4pt]
x \leq s \\[4pt]
-x \leq s \\[4pt]
Ax + Bz = \eta \\[4pt]
z \geq 0 \\[4pt]
t \geq 0.
\end{cases}
\qquad (\text{A.24})
$$

Lemma A.8 *The logarithmic barrier function of the reformulated dual l_p-programming problem $(\mathcal{DL}_p'')$ is 2–self-concordant.*

Proof: Similarly to the proof of Lemma A.7, it can be proved that

$$
-\ln(t_i^{\rho_i} z_k^{-\rho_i+1} - s_i) - \ln t_i,
$$

with $\rho_i \leq 1$, is $(1+\frac{\sqrt{2}}{3}(\rho_i+1))$–self-concordant. The lemma follows now from Lemma A.1 and from $\rho_i \leq 1$. $\qquad\square$

A.6   Other smoothness conditions

Relative Lipschitz Condition

As shown in Jarre [68], if the problem functions $f_i \in C^3$ and fulfil the Relative Lipschitz Condition (2.61) with parameter M, then the associated logarithmic barrier function is $(1+M)$-self-concordant.

Monteiro and Adler's Condition

Monteiro and Adler [112] considered problems with linear equality constraints and convex separable objective function. The objective function $\sum_i f_i(x_i)$ must satisfy the following condition:

There exist real numbers T and p such that for all reals $x > 0$ and $y > 0$ and all $i = 1, \cdots, n$, we have

$$y|f_i'''(y)| \leq T \max\left\{(\frac{x}{y})^p, (\frac{y}{x})^p\right\} f_i''(x).$$

Using Lemma A.2 and substituting $y = x$ in the above condition, it is easy to see that the logarithmic barrier function for such a problem is $(1+\frac{1}{3}T)$-self-concordant.

Scaled Lipschitz Condition

In Zhu [170] and Kortanek and Zhu [84] interior point methods are given and analyzed for problems with equality constraints and convex objective function $f(x)$ which satisfies the so-called Scaled Lipschitz Condition:

Given any γ, $0 < \gamma < 1$, there exists $K > 0$, such that

$$\|X(\nabla f(x + \Delta x) - \nabla f(x) - \nabla^2 f(x)\Delta x)\| \leq K \Delta x^T \nabla^2 f(x)\Delta x, \qquad (A.25)$$

whenever $x > 0$ and $\|X^{-1}\Delta x\| \leq \gamma$.

This condition is also covered by the self-concordance condition if f is three times continuously differentiable in the interior of the feasible domain. More precisely we will show in the next lemma that the corresponding logarithmic barrier function is $(1 + \frac{2}{3}K)$-self-concordant.

Lemma A.9 *Suppose $f(x) \in C^3$ fulfils the Scaled Lipschitz Condition with parameter K. Then the corresponding logarithmic barrier function is $(1 + \frac{2}{3}K)$-self-concordant.*

Proof: It suffices to prove (A.1). Set $h = \Delta x$ as in definition (A.25). First note that

$$\sqrt{\sum_{i=1}^{n} \frac{h_i^2}{x_i^2}} = \|X^{-1}\Delta x\|.$$

Since $f \in C^3$ we may expand ∇f as follows:

$$\nabla f(x + \Delta x) = \nabla f(x) + \nabla^2 f(x)\Delta x + \frac{1}{2}\nabla^3 f(x)[\Delta x, \Delta x, .] + o(\|\Delta x\|^2),$$

where $\nabla^3 f(x)[\Delta x, \Delta x, .]$ is a vector a for which

$$a_i = \sum_{j,k} \frac{\partial^3 f(x)}{\partial x_i\, \partial x_j\, \partial x_k} \Delta x_j\, \Delta x_k.$$

Replacing Δx by $\lambda \Delta x$ in definition (A.25), inserting the above expansion, dividing by λ^2, and taking the limit as λ tends to zero we obtain

$$\|X\nabla^3 f(x)[\Delta x, \Delta x, .]\| \le 2K\Delta x^T \nabla^2 f(x)\Delta x. \qquad (A.26)$$

Giving that

$$\begin{aligned} \nabla^3 f(x)[\Delta x, \Delta x, v] &= \nabla^3 f(x)[\Delta x, \Delta x, .]^T v \\ &= a^T v, \end{aligned}$$

the left–hand side of (A.26) is equal to

$$\max_v \{ \nabla^3 f(x)[\Delta x, \Delta x, v] \ : \ \|X^{-1}v\| = 1 \},$$

or equivalently

$$\max_v \{ a^T v \ : \ \|X^{-1}v\| = 1 \}.$$

In particular, we may choose $v = \Delta x / \|X^{-1}\Delta x\|$ and obtain

$$\nabla^3 f(x)[\Delta x, \Delta x, \Delta x] \le 2K\|X^{-1}\Delta x\|\Delta x^T \nabla^2 f(x)\Delta x,$$

which is exactly relation (A.1). □

Appendix B

General technical lemmas

Lemma B.1 *If G_1, G_2 are symmetric matrices with $|h^T G_1 h| \leq h^T G_2 h$, $\forall h \in \mathbb{R}^n$, then*

$$(h_1^T G_1 h_2)^2 \leq h_1^T G_2 h_1 h_2^T G_2 h_2,$$

$\forall h_1, h_2 \in \mathbb{R}^n$.

Proof: See Jarre [68]. □

Lemma B.2 *If $F \in \mathbb{R}^{n \times n \times n}$ is a symmetric trilinear form, and $G \in \mathbb{R}^{n \times n}$ a symmetric bilinear form, and $\zeta > 0$ is such that $\forall h \in \mathbb{R}^n$*

$$F[h, h, h]^2 \leq \zeta G[h, h]^3,$$

then

$$F[h_1, h_2, h_3]^2 \leq \zeta G[h_1, h_1] G[h_2, h_2] G[h_3, h_3],$$

$\forall h_1, h_2, h_3 \in \mathbb{R}^n$.

Proof: See Nesterov and Nemirovsky [119] and Jarre [68]. □

Lemma B.3 *Let G_1 and G_2 be square matrices of the same size. If G_2 is positive semi-definite and $G_2 + G_1 G_1^T$ is invertible, then the eigenvalues of $G_1^T (G_2 + G_1 G_1^T)^{-1} G_1$ are all smaller than or equal to one. If in addition G_1 is positive definite, then $(G_1 + G_2)^{-1} \preceq G_1^{-1}$.*

Proof: Let $\sigma(P)$ denote the maximal eigenvalue of $P = G_1^T (G_2 + G_1 G_1^T)^{-1} G_1$. It is easy to verify that

$$
\begin{aligned}
P^2 &= G_1^T (G_2 + G_1 G_1^T)^{-1} G_1 G_1^T (G_2 + G_1 G_1^T)^{-1} G_1 \\
&\preceq G_1^T (G_2 + G_1 G_1^T)^{-1} G_1 \\
&= P.
\end{aligned}
$$

Hence, $\sigma(P)^2 = \sigma(P^2) \leq \sigma(P)$. This means that $\sigma(P) \leq 1$, which proves the first part of the lemma. To prove the second part, note that G_1 can be written as RR^T, since G_1 is positive definite. Furthermore,

$$G_1^{-1} - (G_1 + G_2)^{-1} \succeq 0$$

if and only if

$$G := R^T \left(G_1^{-1} - (G_1 + G_2)^{-1} \right) R \succeq 0.$$

Note that

$$G = I - R^T (RR^T + G_2)^{-1} R.$$

Moreover, since G_2 is positive semi–definite it follows from the first part that the eigenvalues of $R^T (RR^T + G_2)^{-1} R$ are smaller than or equal to one. Hence $G \succeq 0$, which proves the second part of the lemma.   □

Lemma B.4 Let $0 \leq \alpha \leq 1$ and $0 \leq \beta < 1$. Then

$$\prod_{i=0}^{\infty} (1 + \alpha \beta^{2^i}) \leq 1 + \frac{\alpha \beta}{1 - \beta}.$$

Proof: It is well–known that

$$\prod_{i=0}^{\infty} (1 + \beta^{2^i}) = 1 + \sum_{i=1}^{\infty} \beta^i.$$

Now, it is easy to verify that

$$\prod_{i=0}^{\infty} (1 + \alpha \beta^{2^i}) = 1 + \sum_{i=1}^{\infty} \alpha^{j_i} \beta^i,$$

where $j_i \geq 1$. Hence it follows that

$$\prod_{i=0}^{\infty} (1 + \alpha \beta^{2^i}) \leq 1 + \alpha \sum_{i=1}^{\infty} \beta^i$$

$$= 1 + \frac{\alpha \beta}{1 - \beta},$$

which proves the lemma.   □

Lemma B.5 For $v \geq 0$ we have: $\ln(1 + v) \leq v - \frac{v^2}{2(1+v)}$.

Proof: First note that $-\ln(1 + v) = \ln(1 - \frac{v}{1+v})$. Now using Karmarkar's [72] well–known inequality we have for $v \geq 0$

$$\ln(1 - \frac{v}{1+v}) \geq -\frac{v}{1+v} - \frac{1}{2} \frac{\left(\frac{v}{1+v}\right)^2}{1 - \frac{v}{1+v}} = -\frac{v}{1+v} - \frac{v^2}{2(1+v)}.$$

This means that

$$\ln(1+v) \leq \frac{v}{1+v} + \frac{v^2}{2(1+v)} = v - \frac{v^2}{2(1+v)}.$$

□

Lemma B.6 *(Anstreicher [3]) Let* $w \in \mathbb{R}$, $0 < w < 1$, *and* $v \in \mathbb{R}$, $v \geq w$. *Then* $|\ln v| \leq \frac{|1-v||\ln w|}{1-w}$.

Proof: Defining

$$\varphi(u) := \begin{cases} \frac{\ln u}{u-1} & \text{if } u \neq 1 \\ 1 & \text{if } u = 1, \end{cases}$$

it is easy to see that $\varphi(u)$ is monotonically decreasing and positive for $u \in (0, \infty)$. Hence

$$\left| \frac{\ln v}{v-1} \right| \leq \left| \frac{\ln w}{w-1} \right| = \frac{|\ln w|}{1-w}.$$

This implies the lemma.

□

Bibliography

[1] Adler, I., Karmarkar, N.K., Resende, M.G.C. and Veiga, G. (1989), An Implementation of Karmarkar's Algorithm for Linear Programming, *Mathematical Programming* 44, 297–335.

[2] Anstreicher, K.M. (1986), A Monotonic Projective Algorithm for Fractional Linear Programming, *Algorithmica* 1, 483–498.

[3] Anstreicher, K.M. (1990), A Standard Form Variant, and Safeguarded Linesearch, for the Modified Karmarkar Algorithm, *Mathematical Programming* 47, 337–351.

[4] Anstreicher, K.M. (1990), On Long Step Path Following and SUMT for Linear and Quadratic Programming, Working Paper, Department of Operations Research, Yale University, New Haven, Connecticut.

[5] Anstreicher, K.M. (1992), Efficient Centering for Linear Programming Interior Point Algorithms, Working Paper, Department of Management Sciences, University of Iowa, Iowa City, Iowa.

[6] Anstreicher, K.M. and Bosch, R.A. (1992), Long Steps in an $O(n^3 L)$ Algorithm for Linear Programming, *Mathematical Programming* 54, 251–265.

[7] Anstreicher, K.M., den Hertog, D., Roos, C. and Terlaky, T. (1990), A Long Step Barrier Method for Convex Quadratic Programming, Report No. 90–53, Faculty of Mathematics and Informatics/Computer Science, Delft University of Technology, Delft, Holland. To Appear in *Algorithmica*.

[8] Bahn, O., Goffin, J.-L., Vial, J.-Ph. and Du Merle, O. (1991), Implementation and Behavior of an Interior Point Cutting Plane Algorithm for Convex Programming: an Application to Geometric Programming, Working Paper, Université de Genève, Genève, Switzerland. To Appear in *Discrete Applied Mathematics*.

[9] Barnes, E.R. (1986), A Variation on Karmarkar's Algorithm for Solving Linear Programming Problems, *Mathematical Programming* 36, 174–182.

[10] Barnes, E.R., Chopra, S. and Jensen, D.L. (1988), A Polynomial Time Version of the Affine Scaling Algorithm, Technical Report, Department of Mathematical Sciences, IBM T.J. Watson Research Center, Yorktown Heights, NY.

[11] Bayer, D.A. and Lagarias, J.C. (1989), The Nonlinear Geometry of Linear Programming I. Affine and Projective Scaling Trajectories II. Legendre Transform Coordinates and Central Trajectories III. Projective Legendre Transform Coordinates and Hilbert Geometry, *Transactions of the American Mathematical Society* 314, 2, 499–581.

[12] Ben Daya, M. and Shetty, C.M. (1990), Polynomial Barrier Function Algorithms for Convex Quadratic Programming, *Arabian Journal for Science and Engineering* 15, 657–670.

[13] Borchers, B. and Mitchell, J.E. (1991), Using an Interior Point Method in a Branch and Bound Algorithm for Integer Programming, R.P.I. Math. Report No. 195, Department of Mathematical Sciences, Rensselaer Polytechnic Institute, Troy, New York.

[14] Bosch, R.A. and Anstreicher, K.M. (1990), On Partial Updating in a Potential Reduction Linear Programming Algorithm of Kojima, Mizuno and Yoshise, Preprint, Department of Operations Research, Yale University, New Haven, Connecticut. To appear in *Algorithmica*.

[15] Carpenter, T.J., Lustig, I.J., Mulvey, J.M. and Shanno, D.F. (1990), A Primal–Dual Interior Point Method for Convex Nonlinear Programs, Rutcor Research Report No. 25–90, Rutgers Center for Operations Research, New Brunswick, New Jersey.

[16] Choi, I.C., Monma, C.L. and Shanno, D.F. (1990), Further Development of a Primal–Dual Interior Point Method, *ORSA Journal on Computing* 2, 4, 304–311.

[17] Dantzig, G.B. and Ye, Y. (1990), A Build–Up Interior Method for Linear Programming: Affine Scaling Form, Technical Report SOL 90–4, Systems Optimization Laboratory, Department of Operations Research, Stanford University, Stanford, California.

[18] De Ghellinck, G. and Vial, J.–Ph. (1986), A Polynomial Newton Method for Linear Programming, *Algorithmica* 1, 425–453.

[19] Den Hertog, D., Jarre, F., Roos, C. and Terlaky, T. (1992), A Sufficient Condition for Self–Concordance with Application to Some Classes of Structured Convex Programming Problems. To Appear in *Mathematical Programming*.

[20] Den Hertog, D., Kaliski, J.A., Roos, C. and Terlaky, T. (1993), A Logarithmic Barrier Cutting Plane Method for Convex Programming, Report No. 93–43, Faculty of Mathematics and Informatics/Computer Science, Delft University of Technology, Delft, Holland. Submitted to *Annals of Operations Research*.

[21] Den Hertog, D. and Roos, C. (1991), A Survey of Search Directions in Interior Point Methods for Linear Programming, *Mathematical Programming* 52, 481–509.

[22] Den Hertog, D., Roos, C. and Terlaky, T. (1991), A Potential–Reduction Variant of Renegar's Short–Step Path–Following Method for Linear Programming, *Linear Algebra and its Applications* 152, 43–68.

[23] Den Hertog, D., Roos, C. and Terlaky, T. (1991), A Polynomial Method of Weighted Centers for Convex Quadratic Programming, *Journal of Information & Optimization Sciences* 12, 2, 187–205.

[24] Den Hertog, D., Roos, C. and Terlaky, T. (1991), Inverse Barrier Methods for Linear Programming, Report No. 91-27, Faculty of Mathematics and Informatics/Computer Science, Delft University of Technology, Delft, Holland. To Appear in *Revue RAIRO-Operations Research*.

[25] Den Hertog, D., Roos, C. and Terlaky, T. (1992), A Build–Up Variant of the Path–Following Method for LP, *Operations Research Letters* 12, 181–186.

[26] Den Hertog, D., Roos, C. and Terlaky, T. (1992), On the Monotonicity of the Dual Objective Along Barrier Paths, *COAL Bulletin* 20, 2–7.

[27] Den Hertog, D., Roos, C. and Terlaky, T. (1992), Adding and Deleting Constraints in the Path–Following Method for LP, SHELL Report AMER.92.001, Koninklijke/Shell–Laboratorium, Amsterdam, Holland.

[28] Den Hertog, D., Roos, C. and Terlaky, T. (1992), A Large–Step Analytic Center Method for a Class of Smooth Convex Programming Problems, *SIAM Journal on Optimization* 2, 55–70.

[29] Den Hertog, D., Roos, C. and Terlaky, T. (1992), On the Classical Logarithmic Barrier Method for a Class of Smooth Convex Programming Problems, *Journal of Optimization Theory and Applications* 73, 1, 1–25.

[30] Den Hertog, D., Roos, C. and Vial, J.-Ph. (1992), A Complexity Reduction for the Long–Step Path–Following Algorithm for Linear Programming, *SIAM Journal on Optimization* 2, 71–87.

[31] Dikin, I.I. (1967), Iterative Solution of Problems of Linear and Quadratic Programming, *Doklady Akademiia Nauk SSSR* 174, 747–748. Translated into English in *Soviet Mathematics Doklady* 8, 674–675.

[32] Dikin, I.I. (1974). On the Convergence of an Iterative Process. *Upravlyaemye Sistemi* 12, 54–60.

[33] Domich, P.D., Boggs, P.T., Rogers, J.E. and Witzgall, C. (1991), Optimizing over Three–dimensional Subspaces in an Interior–point Method for Linear Programming, *Linear Algebra and its Applications* 152, 315–342.

[34] Duffin, R.J., Peterson, E.L. and Zener, C. (1967), *Geometric Programming*, John Wiley & Sons, New York.

[35] Fiacco, A.V. and McCormick, G.P. (1968), *Nonlinear Programming, Sequential Unconstrained Minimization Techniques*, John Wiley & Sons, New York.

[36] Freund, R.M. (1991), Polynomial–Time Algorithms for Linear Programming Based only on Primal Scaling and Projected Gradients of a Potential Function, *Mathematical Programming* 51, 203–222.

[37] Frisch, K.R. (1955), The Logarithmic Potential Method for Solving Linear Programming Problems, Memorandum, Institute of Economics, Oslo, Norway.

[38] Frisch, K.R. (1955), The Logarithmic Potential Method of Convex Programming, Memorandum, Institute of Economics, Oslo, Norway.

[39] Gay, D.M. (1987), A Variant of Karmarkar's Linear Programming Algorithm for Problems in Standard Form, *Mathematical Programming* 37, 81–90.

[40] Gill, P.E., Murray, W., Saunders, M.A., Tomlin, J.A. and Wright, M.H. (1986), On Projected Newton Barrier Methods for Linear Programming and an Equivalence to Karmarkar's Projective Method, *Mathematical Programming* 36, 183–209.

[41] Gill, P.E., Murray, W. and Wright, M.H. (1988), *Practical Optimization*, Seventh Printing, Academic Press, Inc., San Diego, California.

[42] Goffin, J.-L. and Vial, J.-Ph. (1990), Cutting Planes and Column Generation Techniques with the Projective Algorithm, *Journal of Optimization Theory and Applications* 65, 409–429.

[43] Goffin, J.-L. and Vial, J.-Ph. (1990), Short Steps with Karmarkar's Projective Algorithm for Linear Programming, GERAD, Faculty of Management, McGill University, Montréal, Canada.

[44] Goldfarb, D. and Liu, S. (1991), An $O(n^3 L)$ Primal Interior Point Algorithm for Convex Quadratic Programming, *Mathematical Programming* 49, 325–340.

[45] Goldfarb, D. and Todd, M.J. (1989), Linear Programming, in: *Handbooks in Operations Research and Management Science, Optimization*, 1, G.L. Nemhauser, A.H.G. Rinnooy Kan and M.J. Todd, eds., Amsterdam, 141–170.

[46] Golub, G.H. and Van Loan, C.F. (1989), *Matrix Computations*, The Johns Hopkins Press Ltd., London.

[47] Gonzaga, C.C. (1989), An Algorithm for Solving Linear Programming Problems in $O(n^3 L)$ Operations, in: *Progress in Mathematical Programming, Interior Point and Related Methods*, 1–28, N. Megiddo, ed., Springer Verlag, New York.

[48] Gonzaga, C.C. (1989), Conical Projection Algorithms for Linear Programming, *Mathematical Programming* 43, 151–173.

[49] Gonzaga, C.C. (1990), Polynomial Affine Algorithms for Linear Programming, *Mathematical Programming* 49, 7–21.

[50] Gonzaga, C.C. (1990), Convergence of the Large Step Primal Affine–Scaling Algorithm for Primal Non–degenerate Linear Programs, Report ES–230/90, Department of Systems Engineering and Computer Sciences, COPPE–Federal University of Rio de Janeiro, Rio de Janeiro, Brasil.

[51] Gonzaga, C.C. (1991), Large–Steps Path–Following Methods for Linear Programming, Part I: Barrier Function Method, *SIAM Journal on Optimization* 1, 268–279.

[52] Gonzaga, C.C. (1991), Large–Steps Path–Following Methods for Linear Programming, Part II: Potential Reduction Method, *SIAM Journal on Optimization* 1, 280–292.

[53] Gonzaga, C.C. (1991), Search Directions for Interior Linear Programming Methods, *Algorithmica* 6, 153–181.

[54] Gonzaga, C.C. (1991), Interior Point Algorithms for Linear Programming Problems with Inequality Constraints, *Mathematical Programming* 52, 209–225.

[55] Gonzaga, C.C. (1992), Path Following Methods for Linear Programming, *Siam Review* 34, 167–224.

[56] Gonzaga, C.C. and Carlos, L.A. (1990), A Primal Affine–Scaling Algorithm for Linearly Constrained Convex Programs, Report ES–238/90, Department of Systems Engineering and Computer Sciences, COPPE–Federal University of Rio de Janeiro, Rio de Janeiro, Brasil.

[57] Güler, O., Roos, C., Terlaky, T. and Vial J.–Ph. (1992), Interior Point Approach to the Theory of Linear Programming, Technical Report 1992.3, Université de Genève, Département d'Economie Commerciale et Industrielle, Genève, Switzerland.

[58] Hamala, M. (1976), Quasibarrier Methods for Convex Programming, *Survey of Mathematical Programming* 1, 465–477, Proceedings of the IX–th International Symposium on Mathematical Programming (Budapest, 1976), North–Holland.

[59] Han, C.–G., Pardalos, P.M. and Ye, Y. (1991), On Interior–Point Algorithms for Some Entropy Optimization Problems, Working Paper, Computer Science Department, The Pennsylvania State University, University Park, Pennsylvania.

[60] Huard, P. (1967), Resolution of Mathematical Programming with Nonlinear Constraints by the Method of Centres, In: *Nonlinear Programming*, J. Abadie ed., North–Holland Publishing Company, Amsterdam, Holland, 207–219.

[61] Imai, H. (1988), On the Convexity of the Multiplicative Version of Karmarkar's Potential Function, *Mathematical Programming* 40, 29–32.

[62] Iri, M. (1991), A Proof of the Polynomiality of the Iri–Imai Method, Report METRO 91–08, Mathematical Engineering Section, Faculty of Engineering, University of Tokyo, Tokyo, Japan.

[63] Iri, M. and Imai, H. (1986), A Mutiplicative Barrier Function Method for Linear Programming, *Algorithmica* 1, 455–482.

[64] Jansen, B., Roos, C. and Terlaky, T. (1992), An Interior Point Approach to Postoptimal and Parametric Analysis in Linear Programming, Report No. 92–21, Faculty of Mathematics and Informatics/Computer Science, Delft University of Technology, Delft, Holland.

[65] Jansen, B., Roos, C., Terlaky, T. and Vial, J.-Ph. (1992), Primal–Dual Algorithms For Linear Programming Based on the Logarithmic Barrier Method, Report No. 92–104, Faculty of Mathematics and Informatics/Computer Science, Delft University of Technology, Delft, Holland. To Appear in *Journal of Optimization Theory and Applications* 83.

[66] Jarre, F. (1989), On the Method of Analytic Centers for Solving Smooth Convex Programs, In: S. Dolecki, ed., *Lecture Notes in Mathematics* 1405, Optimization, Springer, 69–85.

[67] Jarre, F. (1991), On the Convergence of the Method of Analytic Centers When Applied to Convex Quadratic Programs, *Mathematical Programming* 49, 341–358.

[68] Jarre, F. (1992), Interior–point Methods for Convex Programming, *Applied Mathematics & Optimization* 26, 287–311.

[69] Jarre, F. and Saunders, M. (1991), Practical Aspects of an Interior–Point Method for Convex Programming, Technical Report SOL 91-9, Systems Optimization Laboratory, Department of Operations Research, Stanford University, Stanford, California.

[70] Kaliski, J.A. and Ye, Y. (1991), A Decomposition Variant of the Potential Reduction Algorithm for Linear Programming, Working Paper 91–11, Department of Management Sciences, The University of Iowa, Iowa City, Iowa.

[71] Kapoor, S. and Vaidya, P.M. (1986), Fast Algorithms for Convex Quadratic Programming and Multicommodity Flows, *Proceedings of the 18th Annual ACM Symposium on the Theory of Computing*, 147–159.

[72] Karmarkar, N.K. (1984), A New Polynomial–Time Algorithm for Linear Programming, *Combinatorica* 4, 373–395.

[73] Khachian, L.G. (1979), A Polynomial Algorithm in Linear Programming, *Doklady Akademiia Nauk SSSR* 244, 1093–1096. Translated into English in *Soviet Mathematics Doklady* 20, 191–194.

[74] Kieviet, C.J. (1989), *Toen Dik Trom een Jongen was*, Edited by R. van Steenbergen, Rebo–Productions, Lisse, Holland.

[75] Klee, V. and Minty, G.J. (1972), How Good is the Simplex Algorithm?, in: *Inequalities III*, O. Shisha, ed., Academic Press, New York, 159–175.

[76] Kojima, M., Megiddo, N., and Mizuno, S. (1991), A Primal–Dual Exterior Point Algorithm for Linear Programming, Research Report RJ 8500 (76721), IBM Almaden Research Division, San Jose, California.

[77] Kojima, M., Megiddo, N., Noma, T. and Yoshise, A. (1991), A Unified Approach to Interior Point Algorithms for Linear Complementarity Problems, *Lecture Notes in Computer Science* 538.

[78] Kojima, M., Mizuno, S. and Yoshise, A. (1989), A Primal–Dual Interior Point Algorithm for Linear Programming, in: *Progress in Mathematical Programming, Interior Point and Related Methods*, 29–47, N. Megiddo ed., Springer Verlag, New York.

[79] Kojima, M., Mizuno, S. and Yoshise, A. (1989), A Polynomial Time Algorithm for a Class of Linear Complementarity Problems, *Mathematical Programming* 44, 1–26.

[80] Kojima, M., Mizuno, S. and Yoshise, A. (1991), An $O(\sqrt{n}L)$ Iteration Potential Reduction Algorithm for Linear Complementarity Problems, *Mathematical Programming* 50, 331–342.

[81] Kortanek, K.O. and No, H. (1992), A Second Order Affine Scaling Algorithm for the Geometric Programming Dual with Logarithmic Barrier, *Optimization* 23, 501–507.

[82] Kortanek, K.O., Potra, F. and Ye, Y. (1991), On Some Efficient Interior Point Methods for Nonlinear Convex Programming, *Linear Algebra and its Applications* 152, 169–189.

[83] Kortanek, K.O. and Zhu, J. (1988), New Purification Algorithms for Linear Programming, *Naval Research Logistics Quarterly* 35, 571–583.

[84] Kortanek, K.O. and Zhu, J. (1993), A Polynomial Barrier Algorithm for Linearly Constrained Convex Programming Problems, *Mathematics of Operations Research* 18, 116–127.

[85] Kranich, E. (1991), Interior Point Methods for Mathematical Programming: a Bibliography, Diskussionsbeitrag Nr. 171, Fern Universität Hagen, Hagen, Germany. (An Updated Version is Available on Netlib.)

[86] Lootsma, F.A. (1970), Boundary Properties of Penalty Functions for Constrained Minimization, Ph.D. Thesis, Eindhoven, The Netherlands.

[87] Lustig, I.J., Marsten, R.E. and Shanno, D.F. (1992), On Implementing Mehrotra's Predictor–Corrector Interior Point Method for Linear Programming, *SIAM Journal on Optimization* 2, 435–449.

[88] Lustig, I.J., Marsten, R.E. and Shanno, D.F. (1991), Computational Experience With a Primal–Dual Interior Point Method for Linear Programming, *Linear Algebra and Its Applications* 152, 191–222.

[89] Lustig, I.J., Marsten, R.E. and Shanno, D.F. (1991), Interior Method VS Simplex Method: Beyond Netlib, *COAL Newsletter* 19, 41–44.

[90] Marsten, R.E., Saltzman, M.J., Shanno, D.F., Pierce, G.S. and Ballintijn, J.F. (1989), Implementation of a Dual Affine Interior Point Algorithm for Linear Programming, *ORSA Journal on Computing* 1, 4, 287–297.

[91] Marsten, R.E., Subramanian, R., Saltzman, M.J., Lustig, I.J. and Shanno, D.F. (1990), Interior Point Methods for Linear Programming: Just Call Newton, Lagrange, and Fiacco and McCormick!, *Interfaces* 20, 105–116.

[92] McCormick, G.P. (1989), The Projective SUMT Method for Convex Programming, *Mathematics of Operations Research* 14, 2, 203–223.

[93] McCormick, G.P. (1991), The Superlinear Convergence of a Nonlinear Primal–Dual Algorithm, Research Report GWU/OR/Serial T-550/91, The George Washington University, School of Engineering and Applied Science, Department of Operations Research, Washington.

[94] McLinden, L. (1980), An Analogue of Moreau's Proximation Theorem, With Application to the Nonlinear Complementarity Problem, *Pacific Journal of Mathematics* 88, 1, 101–161.

[95] McShane, K.A., Monma, C.L. and Shanno, D.F. (1989), An Implementation of a Primal–Dual Interior Point Method for Linear Programming, *ORSA Journal on Computing* 1, 70–83.

[96] Megiddo, N. (1989), Pathways to the Optimal Set in Linear Programming, in: *Progress in Mathematical Programming, Interior Point and Related Methods*, 131–158, N. Megiddo ed., Springer Verlag, New York.

[97] Megiddo, N. and Shub, M. (1989), Boundary Behavior of Interior Point Algorithms in Linear Programming, *Mathematics of Operations Research* 14, 1, 97–146.

[98] Mehrotra, S. (1992), On the Implementation of a (Primal–Dual) Interior Point Method, *SIAM Journal on Optimization* 2, 575–601.

[99] Mehrotra, S. and Sun, J. (1990), An Interior Point Algorithm for Solving Smooth Convex Programs Based on Newton's Method, *Mathematical Developments Arising from Linear Programming*, J.C. Lagarias and M.J. Todd, eds., *Contemporary Mathematics* 114, 265–284.

[100] Mehrotra, S. and Sun, J. (1991), On Computing the Center of a Convex Quadratically Constrained Set, *Mathematical Programming* 50, 81–89.

[101] Minoux, M. (1986), *Mathematical Programming, Theory and Algorithms*, John Wiley & Sons, New York.

[102] Mitchell, J.E. (1988), Karmarkar's Algorithm and Combinatorial Optimization Problems, Ph.D. Thesis, Cornell University, Ithaca.

[103] Mitchell, J.E. and Todd, M.J. (1989), On the Relationship between the Search Directions in the Affine and Projective Variants of Karmarkar's Linear Programming Algorithm, in: *Contributions to Operations Research and Economics*, B. Cornet and H. Tulkens, eds., MIT Press, Cambridge, MA, 237–250.

[104] Mizuno, S. (1990), An $O(n^3L)$ Algorithm Using a Sequence for a Linear Complementarity Problem, *Journal of the Operations Research Society of Japan* 33, 66–75.

[105] Mizuno, S. and Nagasawa, A. (1992), A Primal–Dual Affine Scaling Potential Reduction Algorithm for Linear Programming, Research Memorandum No. 427, Department of Prediction and Control, The Institute of Statistical Mathematics, Tokyo, Japan.

[106] Mizuno, S. and Todd, M.J. (1991), An $O(n^3L)$ Adaptive Path Following Algorithm for a Linear Complementarity Problem, *Mathematical Programming* 52, 587–592.

[107] Monma, C.L. and Morton, A.J. (1987), Computational Experience with a Dual Affine Variant of Karmarkar's Method for Linear Programming, *Operations Research Letters* 6, 261–267.

[108] Monteiro, R.D.C. (1991), The Global Convergence of a Class of Primal Potential Reduction Algorithms for Convex Programming, Working Paper No. 91-024, University of Arizona, Systems and Industrial Engineering Department, Tucson.

[109] Monteiro, R.D.C. (1991), A Globally Convergent Interior Point Algorithm for Convex Programming, Working Paper No 91-021, University of Arizona, Systems and Industrial Engineering Department, Tucson.

[110] Monteiro, R.D.C. and Adler, I. (1989), Interior Path Following Primal–Dual Algorithms, Part I: Linear Programming, *Mathematical Programming* 44, 27–41.

[111] Monteiro, R.D.C. and Adler, I. (1989), Interior Path Following Primal–Dual Algorithms, Part II: Convex Quadratic Programming, *Mathematical Programming* 44, 43–66.

[112] Monteiro, R.D.C. and Adler, I. (1990), An Extension of Karmarkar–type Algorithms to a Class of Convex Separable Programming Problems with Global Rate of Convergence, *Mathematics of Operations Research* 15, 408–422.

[113] Monteiro, R.D.C., Adler, I. and Resende, M.G.C. (1990), A Polynomial–Time Primal–Dual Affine Scaling Algorithm for Linear and Convex Quadratic Programming and its Power Series Extension, *Mathematics of Operations Research* 15, 191–214.

[114] Morin, T.L. and Trafalis, T.B. (1989), A Polynomial–Time Algorithm for Finding an Efficient Face of a Polyhedron, Technical Report, Purdue University, West Lafayette, Indiana.

[115] Murray, W. and Wright, M.H. (1992), Line Search Procedures for the Logarithmic Barrier Function, Numerical Analysis Manuscript 92–01, AT&T Bell Laboratories, Murray Hill, New Jersey.

[116] Murty, K.G. (1988), *Linear Complementarity, Linear and Nonlinear Programming*, Heldermann Verlag Berlin, Germany.

[117] Mylander, W.C., Holmes, R.L. and McCormick, G.P. (1971), A Guide to SUMT-Version 4: the Computer Program Implementing the Sequential Unconstrained Minimization Technique for Nonlinear Programming, Research Paper RAC–P–63, Research Analysis Corporation, McLean, VA.

[118] Nash, S.G. and Sofer, A. (1993), A Barrier Method for Large–Scale Constrained Optimization, *ORSA Journal on Computing* 5, 40–53.

[119] Nesterov, Y.E. and Nemirovsky, A.S. (1989), Self–Concordant Functions and Polynomial Time Methods in Convex Programming, Report, Central Economical and Mathematical Institute, USSR Academy of Science, Moscow, USSR.

[120] Papadimitriou, C.H. and Steiglitz, K. (1982), *Combinatorial Optimization: Algorithms and Complexity*, Prentice–Hall, Englewood Cliffs, New Jersey.

[121] Parisot, G.R. (1961), Résolution Numérique Approchée du Problème de Programmation Linèaire par Application de la Programmation Logarithmique, *Revue Française Recherche Opérationelle* 20, 227-259.

[122] Peterson, E.L., and Ecker, J.G. (1970), Geometric Programming: Duality in Quadratic Programming and l_p Approximation I, in: H.W. Kuhn and A.W. Tucker (eds.), *Proceedings of the International Symposium of Mathematical Programming*, Princeton University Press, New Jersey.

[123] Peterson, E.L., and Ecker, J.G. (1967), Geometric Programming: Duality in Quadratic Programming and l_p Approximation II, *SIAM Journal on Applied Mathematics* 13, 317–340.

[124] Peterson, E.L., and Ecker, J.G. (1970), Geometric Programming: Duality in Quadratic Programming and l_p Approximation III, *Journal of Mathematical Analysis and Applications* 29, 365–383.

[125] Polak, E. (1971), *Computational Methods in Optimization*, Academic Press, New York, New York.

[126] Ponceleón, D.B. (1991), Barrier Methods for Large–Scale Quadratic Programming, Ph.D. Thesis, Technical Report SOL 91–2, Systems Optimization Laboratory, Department of Operations Research, Stanford University, Stanford, California.

[127] Renegar, J. (1988), A Polynomial–Time Algorithm, Based on Newton's Method, for Linear Programming, *Mathematical Programming* 40, 59–93.

[128] Roos, C. (1989), New Trajectory–Following Polynomial–Time Algorithm for Linear Programming Problems, *Journal of Optimization Theory and Applications* 63, 3, 433–458.

[129] Roos, C. (1990), A Projective Variant of the Approximate Center Method for Linear Programming, Report No. 90–83, Faculty of Mathematics and Informatics/Computer Science, Delft University of Technology, Delft, Holland.

[130] Roos, C. and Den Hertog, D. (1989), A Polynomial Method of Approximate Weighted Centers for Linear Programming, Report No. 89–13, Faculty of Mathematics and Informatics/Computer Science, Delft University of Technology, Delft, Holland.

[131] Roos, C. and Vial, J.–Ph. (1988), Analytic Centers in Linear Programming, Report No. 88–74, Faculty of Mathematics and Informatics/Computer Science, Delft University of Technology, Delft, Holland.

[132] Roos, C. and Vial, J.–Ph. (1990), Long Steps with the Logarithmic Penalty Barrier Function in Linear Programming, in *Economic Decision–Making: Games, Economics and Optimization*, dedicated to Jacques H. Drèze, J. Gabszevwicz, J.–F. Richard and L. Wolsey, eds., Elsevier Science Publisher B.V., 433–441.

[133] Roos, C. and Vial, J.–Ph. (1992), A Polynomial Method of Approximate Centers for Linear Programming, *Mathematical Programming* 54, 295–305.

[134] Shor, N. (1970), Utilization of the Operation of Space Dilatation in the Minimization of Convex Functions, *Kibernetica* 1, 6–12. English translation: *Cybernetics* 6, 7–15.

[135] Sonnevend, G. (1986), An "Analytical Centre" for Polyhedrons and New Classes of Global Algorithms for Linear (Smooth, Convex) Programming, *Lecture Notes in Control and Information Sciences* 84, 866–878.

[136] Tanabe, K. (1987), Centered Newton Method for Mathematical Programming, *System Modelling and Optimization*, Iri, M., and Yajima, K., eds., 197–206 (Proceedings of the 13th IFIP Conference, Tokyo, Aug. 31–Sept. 4, 1987).

[137] Terlaky, T. (1985), On l_p programming, *European Journal of Operational Research* 22, 70–100.

[138] Todd, M.J. (1989), Recent Developments and New Directions in Linear Programming, in: *Mathematical Programming: Recent Developments and Applications*, M. Iri and K. Tanabe, eds., Kluwer Academic Press, Dordrecht, Holland, 109–157.

[139] Todd, M.J. and Burrell, B.P. (1986), An Extension of Karmarkar's Algorithm for Linear Programming Using Dual Variables, *Algorithmica* 1, 409–424.

[140] Todd, M.J. and Ye, Y. (1990), A Centered Projective Algorithm for Linear Programming, *Mathematics of Operations Research* 15, 508–529.

[141] Tomlin, J.A. (1987), An Experimental Approach to Karmarkar's Projective Method for Linear Programming, *Mathematical Programming Study* 31, 175–191.

[142] Tone, K. (1991), An Active–Set Strategy in Interior Point Method for Linear Programming, Working Paper, Graduate School of Policy Science, Saitama University, Urawa, Saitama 338, Japan.

[143] Tseng, P. and Luo, Z. –Q. (1989), On the Convergence of the Affine–scaling Algorithm, *Preprint CICS–P–169,* M.I.T., Cambridge, Massachusetts.

[144] Tsuchiya, T. (1991), Global Convergence of the Affine Scaling Methods for Degenerate Linear Programming Problems, *Mathematical Programming* 52, 377–404.

[145] Tsuchiya, T. (1991), Global Convergence of the Affine Scaling Algorithm for the Primal Degenerate Strictly Convex Quadratic Programming Problems, Research Memorandum No. 417, The Institute of Statistical Mathematics, Tokyo, Japan.

[146] Tsuchiya, T. (1991), Quadratic Convergence of Iri and Imai's Algorithm for Degenerate Linear Programming Problems, Research Memorandum No. 412, The Institute of Statistical Mathematics, Tokyo, Japan. To Appear in *Journal of Optimization Theory and Applications*.

[147] Tsuchiya, T. (1992), Global Convergence Property of the Affine Scaling Methods for Primal Degenerate Linear Programming Problems, *Mathematics of Operations Research* 17, 527–557.

[148] Tsuchiya, T. and Muramatsu, M. (1992), Global Convergence of a Long–Step Affine Scaling Algorithm for Degenerate Linear Programming Problems, Research Memorandum No. 423, The Institute of Statistical Mathematics, Tokyo, Japan.

[149] Vaidya, P.M. (1990), An Algorithm for Linear Programming which Requires $O(((m+n)n^2+(m+n)^{1.5}n)L)$ Arithmetic Operations, *Mathematical Programming* 47, 175–201.

[150] Vanderbei, R.J. and Lagarias, J.C. (1990), I.I. Dikin's Convergence Result for the Affine–Scaling Algorithm, *Mathematical Developments Arising from Linear Programming*, J.C. Lagarias and M.J. Todd, eds., *Contemporary Mathematics* 114, 109–119.

[151] Vanderbei, R.J., Meketon, M.S. and Freedman, B.A. (1986), A Modification of Karmarkar's Linear Programming Algorithm, *Algorithmica* 1, 395–407.

[152] Wolfe, Ph. (1961), A Duality Theorem for Nonlinear Programming, *Quarterly of Applied Mathematics* 19, 239–244.

[153] Wright, M.H. (1991), Determining Subspace Information from the Hessian of a Barrier Function, Manuscript 92–02, AT&T Bell Laboratories, Murray Hill, New Jersey.

[154] Wright, M.H. (1992), Interior Methods for Constrained Optimization, *Acta Numerica*, Cambridge University Press, New York, 341–407.

[155] Xiao, D. and Goldfarb, D. (1990), A Path–following Projective Interior Point Method for Linear Programming, Working Paper, Department of Industrial Engineering and Operations Research, Columbia University, New York, NY.

[156] Yamashita, H. (1986), A Polynomially and Quadratically Convergent Method for Linear Programming, Working Paper, Mathematical Systems Institute, Inc., Tokyo, Japan.

[157] Yamashita, H. (1991), A Class of Primal–Dual Method for Constrained Optimization, Working Paper, Mathematical Systems Institute, Inc., Tokyo, Japan.

[158] Ye, Y. (1987), Further Development on the Interior Algorithm for Convex Quadratic Programming, Department of Engineering–Economic Systems, Stanford University, Stanford, California.

[159] Ye, Y. (1989), Eliminating Columns and Rows in Potential Reduction and Path–Following Algorithms for Linear Programming, Working Paper Series No. 89–7, Department of Management Sciences, The University of Iowa, Iowa City, Iowa.

[160] Ye, Y. (1990), A Class of Projective Transformations for Linear Programming, *SIAM Journal on Computing* 19, 457–466.

[161] Ye, Y. (1990), The "Build–Down" Scheme for Linear Programming, *Mathematical Programming* 46, 61–72.

[162] Ye, Y. (1991), An $O(n^3L)$ Potential Reduction Algorithm for Linear Programming, *Mathematical Programming* 50, 239–258.

[163] Ye, Y. (1992), A Potential Reduction Algorithm Allowing Column Generation, *SIAM Journal on Optimization* 2, 7–20.

[164] Ye, Y. and Kojima, M. (1987), Recovering Optimal Dual Solutions in Karmarkar's Polynomial Algorithm for Linear Programming, *Mathematical Programming* 39, 305–317.

[165] Ye, Y. and Pardalos, P. (1991), A Class of Linear Complementarity Problems Solvable in Polynomial Time, *Linear Algebra and Its Applications* 152, 1–9.

[166] Ye, Y. and Potra, F. (1990), An Interior–Point Algorithm for Solving Entropy Optimization Problems with Globally Linear and Locally Quadratic Convergence Rate, Working Paper Series No. 90–22, Department of Management Sciences, The University of Iowa, Iowa City, Iowa. To Appear in *SIAM Journal on Optimization*.

[167] Ye, Y. and Tse, E. (1989), An Extension of Karmarkar's Projective Algorithm for Convex Quadratic Programming, *Mathematical Programming* 44, 157–179.

[168] Yudin, D. and Nemirovsky, A.S. (1976), Informational Complexity and Efficient Methods for the Solution of Convex Extremal Problems (in Russian), *Ékonomika i Mathematicheskie Metody* 12, 357–369. English translation: *Matekon* 13(2), 3–25.

[169] Zhang, S. (1989), On the Convergence Property of Iri–Imai's Method for Linear Programming, Report 8917/A, Econometric Institute, Erasmus University, Rotterdam, The Netherlands.

[170] Zhu, J. (1992), A Path Following Algorithm for a Class of Convex Programming Problems, *Zeitschrift für Operations Research – Methods and Models of Operations Research* 36, 359–377.

Index

Other *Mathematics and Its Applications* titles of interest:

B.S. Razumikhin: *Physical Models and Equilibrium Methods in Programming and Economics.* 1984, 368 pp.   ISBN 90-277-1644-7

N.K. Bose (ed.): *Multidimensional Systems Theory. Progress, Directions and Open Problems in Multidimensional Systems.* 1985, 280 pp.   ISBN 90-277-1764-8

J. Szep and F. Forgo: *Introduction to the Theory of Games.* 1985, 412 pp.
                                                                ISBN 90-277-1404-5

V. Komkov: *Variational Principles of Continuum Mechanics with Engineering Applications. Volume 1: Critical Points Theory.* 1986, 398 pp.
                                                                ISBN 90-277-2157-2

V. Barbu and Th. Precupanu: *Convexity and Optimization in Banach Spaces.* 1986, 416 pp.   ISBN 90-277-1761-3

M. Fliess and M. Hazewinkel (eds.): *Algebraic and Geometric Methods in Non-linear Control Theory.* 1986, 658 pp.   ISBN 90-277-2286-2

P.J.M. van Laarhoven and E.H.L. Aarts: *Simulated Annealing: Theory and Applications.* 1987, 198 pp.   ISBN 90-277-2513-6

B.S. Razumikhin: *Classical Principles and Optimization Problems.* 1987, 528 pp.
                                                                ISBN 90-277-2605-1

S. Rolewicz: *Functional Analysis and Control Theory. Linear Systems.* 1987, 544 pp.   ISBN 90-277-2186-6

V. Komkov: *Variational Principles of Continuum Mechanics with Engineering Applications. Volume 2: Introduction to Optimal Design Theory.* 1988, 288 pp.
                                                                ISBN 90-277-2639-6

A.A. Pervozvanskii and V.G. Gaitsgori: *Theory of Suboptimal Decisions. Decomposition and Aggregation.* 1988, 404 pp.   out of print, ISBN 90-277-2401-6

J. Mockus: *Bayesian Approach to Global Optimization. Theory and Applications.* 1989, 272 pp.   ISBN 0-7923-0115-3

Du Dingzhu and Hu Guoding (eds.): *Combinatorics, Computing and Complexity.* 1989, 248 pp.   ISBN 0-7923-0308-3

M. Iri and K. Tanabe: *Mathematical Programming. Recent Developments and Applications.* 1989, 392 pp.   ISBN 0-7923-0490-X

A.T. Fomenko: *Variational Principles in Topology. Multidimensional Minimal Surface Theory.* 1990, 388 pp.   ISBN 0-7923-0230-3

A.G. Butkovskiy and Yu.I. Samoilenko: *Control of Quantum-Mechanical Processes and Systems.* 1990, 246 pp.   ISBN 0-7923-0689-9

A.V. Gheorghe: *Decision Processes in Dynamic Probabilistic Systems.* 1990, 372 pp.   ISBN 0-7923-0544-2

Other *Mathematics and Its Applications* titles of interest:

A.G. Butkovskiy: *Phase Portraits of Control Dynamical Systems*. 1991, 180 pp.
ISBN 0-7923-1057-8

A.A. Zhigljavsky: *Theory of the Global Random Search*. 1991, 360 pp.
ISBN 0-7923-1122-1

G. Ruhe: *Algorithmic Aspects of Flows in Networks*. 1991, 220 pp.
ISBN 0-7923-1151-5

S. Walukuwiecz: *Integer Programming*. 1991, 196 pp.   ISBN 0-7923-0726-7

M. Kisielewicz: *Differential Inclusions and Optimal Control*. 1991, 320 pp.
ISBN 0-7923-0675-9

J. Klamka: *Controllability of Dynamical Systems*. 1991, 260 pp.
ISBN 0-7923-0822-0

V.N. Fomin: *Discrete Linear Control Systems*. 1991, 302 pp.   ISBN 0-7923-1248-1

L. Xiao-Xin: *Absolute Stability of Nonlinear Control Systems*. 1992, 180 pp.
ISBN 0-7923-1988-5

A. Halanay and V. Rasvan: *Applications of Liapunov Methods in Stability*. 1993,
238 pp.   ISBN 0-7923-2120-0

D. den Hertog: *Interior Point Approach to Linear, Quadratic and Convex Programming*. 1994, 208 pp.   ISBN 0-7923-2734-9